$ 14.80

Walt Freiberg

Graduate Texts in Mathematics
51

Editorial Board

F. W. Gehring

P. R. Halmos
Managing Editor

C. C. Moore

Wilhelm Klingenberg

A Course in Differential Geometry

Translated by David Hoffman

Springer-Verlag
New York   Heidelberg   Berlin

Wilhelm Klingenberg
Mathematisches Institut der
Universität Bonn
5300 Bonn
Wegelerstr. 10
West Germany

David Hoffman
Department of Mathematics
Graduate Research Center
University of Massachusetts
Amherst, MA 01003
USA

Editorial Board

P. R. Halmos
Managing Editor
Department of Mathematics
University of California
Santa Barbara, CA 93106
USA

F. W. Gehring
Department of Mathematics
University of Michigan
Ann Arbor, Michigan 48104
USA

C. C. Moore
Department of Mathematics
University of California
Berkeley, CA 94720
USA

AMS Subject Classification: 53-01

Library of Congress Cataloging in Publication Data
Klingenberg, Wilhelm, 1924–
   A course in differential geometry.

   (Graduate texts in mathematics; 51)
   Translation of Eine Vorlesung über Differentialgeometrie.
   Bibliography: p.
   Includes index.
   1. Geometry, Differential. I. Title. II. Series.
QA641.K5813   516'.36   77-4475

All rights reserved.

No part of this book may be translated or reproduced in any form without written permission from Springer-Verlag.

© 1978 by Springer-Verlag, New York Inc.

Printed in the United States of America.

9 8 7 6 5 4 3 2 1

ISBN 0-387-90255-4   Springer-Verlag New York
ISBN 3-540-90255-4   Springer-Verlag Berlin Heidelberg

Dedicated to

Shiing-shen Chern

Preface to the English Edition

This English edition could serve as a text for a first year graduate course on differential geometry, as did for a long time the Chicago Notes of Chern mentioned in the Preface to the German Edition. Suitable references for ordinary differential equations are Hurewicz, W. *Lectures on ordinary differential equations*. MIT Press, Cambridge, Mass., 1958, and for the topology of surfaces: Massey, *Algebraic Topology*, Springer-Verlag, New York, 1977.

Upon David Hoffman fell the difficult task of transforming the tightly constructed German text into one which would mesh well with the more relaxed format of the Graduate Texts in Mathematics series. There are some elaborations and several new figures have been added. I trust that the merits of the German edition have survived whereas at the same time the efforts of David helped to elucidate the general conception of the Course where we tried to put Geometry before Formalism without giving up mathematical rigour.

I wish to thank David for his work and his enthusiasm during the whole period of our collaboration. At the same time I would like to commend the editors of Springer-Verlag for their patience and good advice.

Bonn *Wilhelm Klingenberg*
June, 1977

From the Preface to the German Edition

This book has its origins in a one-semester course in differential geometry which I have given many times at Göttingen, Mainz, and Bonn.

It is my intention that these lectures should offer an introduction to the classical differential geometry of curves and surfaces, suitable for students in their middle semester who have mastered the introductory courses. A course such as this would be an alternative to other middle semester courses such as complex function theory, abstract algebra, or algebraic topology.

For the most part, these lectures assume nothing more than a knowledge of basic analysis, real linear algebra, and euclidean geometry. It is only in the last chapters that a familiarity with the topology of compact surfaces would be useful. Nothing is used that cannot be found in Seifert and Threlfall's classic textbook of topology.

For a summary of the contents of these lectures, I refer the reader to the table of contents. Of course it was necessary to make a selection from the profusion of material that could be presented at this level. For me it was clear that the preferred topics were precisely those which contributed to an understanding of two-dimensional Riemannian geometry. Nonetheless, I think that my lectures provide a useful basis for the understanding of all the areas of differential geometry.

The structure of these lectures, including the organization of some of the proofs, has been greatly influenced by S. S. Chern's lecture notes entitled "Differential Geometry," published in Chicago in 1954. Chern, in turn, was influenced by W. Blaschke's "Vorlesungen über Differentialgeometrie." Chern had studied with Blaschke in Hamburg between 1934 and 1936, and, nearly twenty years later, it was Blaschke who gave me strong support in my career as a differential geometer.

So as I take the privilege of dedicating this book to Shiing-shen Chern, I would at the same time desire to honor the memory of W. Blaschke.

Bonn-Röttgen Wilhelm Klingenberg
January 1, 1972

Contents

Chapter 0
Calculus in Euclidean Space — 1

0.1 Euclidean Space — 1
0.2 The Topology of Euclidean Space — 2
0.3 Differentiation in $\mathbb{R}^n$ — 3
0.4 Tangent Space — 5
0.5 Local Behavior of Differentiable Functions (Injective and Surjective Functions) — 6

Chapter 1
Curves — 8

1.1 Definitions — 8
1.2 The Frenet Frame — 10
1.3 The Frenet Equations — 11
1.4 Plane Curves; Local Theory — 15
1.5 Space Curves — 17
1.6 Exercises — 20

Chapter 2
Plane Curves: Global Theory — 21

2.1 The Rotation Number — 21
2.2 The Umlaufsatz — 24
2.3 Convex Curves — 27
2.4 Exercises and Some Further Results — 29

Chapter 3
Surfaces: Local Theory — 33

- 3.1 Definitions — 33
- 3.2 The First Fundamental Form — 35
- 3.3 The Second Fundamental Form — 38
- 3.4 Curves on Surfaces — 43
- 3.5 Principal Curvature, Gauss Curvature, and Mean Curvature — 45
- 3.6 Normal Form for a Surface, Special Coordinates — 49
- 3.7 Special Surfaces, Developable Surfaces — 54
- 3.8 The Gauss and Codazzi–Mainardi Equations — 61
- 3.9 Exercises and Some Further Results — 66

Chapter 4
Intrinsic Geometry of Surfaces: Local Theory — 73

- 4.1 Vector Fields and Covariant Differentiation — 74
- 4.2 Parallel Translation — 76
- 4.3 Geodesics — 78
- 4.4 Surfaces of Constant Curvature — 83
- 4.5 Examples and Exercises — 87

Chapter 5
Two-dimensional Riemannian Geometry — 89

- 5.1 Local Riemannian Geometry — 90
- 5.2 The Tangent Bundle and the Exponential Map — 95
- 5.3 Geodesic Polar Coordinates — 99
- 5.4 Jacobi Fields — 102
- 5.5 Manifolds — 105
- 5.6 Differential Forms — 111
- 5.7 Exercises and Some Further Results — 119

Chapter 6
The Global Geometry of Surfaces — 123

- 6.1 Surfaces in Euclidean Space — 123
- 6.2 Ovaloids — 129
- 6.3 The Gauss–Bonnet Theorem — 138
- 6.4 Completeness — 144
- 6.5 Conjugate Points and Curvature — 148
- 6.6 Curvature and the Global Geometry of a Surface — 152
- 6.7 Closed Geodesics and the Fundamental Group — 156
- 6.8 Exercises and Some Further Results — 161

References — 167

Index — 171

Index of Symbols — 177

Calculus in Euclidean Space 0

We will start with a brief outline of the essential facts about $\mathbb{R}^n$ and the vector calculus.[1] The reader familiar with this subject may wish to begin with Chapter 1, using this chapter as the need arises.

0.1 Euclidean Space

As usual, $\mathbb{R}^n$ is the vector space of all real n-tuples $x = (x^1, \ldots, x^n)$. The scalar product of two elements x, y in $\mathbb{R}^n$ is given by the formula

$$x \cdot y := \sum_i x^i y^i.$$

We will write $x \cdot x = x^2$ and $\sqrt{x^2} = |x|$. The real number $|x|$ is called the *length* or the *norm* of x. The *Schwarz inequality*,

$$(x \cdot y)^2 \leq |x|^2 |y|^2, \qquad |x|^2 = x^2,$$

is satisfied by the scalar product and from it is derived the *triangle inequality*:

$$|x + y| \leq |x| + |y| \quad \text{for all } x, y \in \mathbb{R}^n.$$

The distinguished basis of $\mathbb{R}^n$ will be denoted by (e_i), $1 \leq i \leq n$. The vector e_i is the n-tuple with 1 in the ith place and 0 in all the other places.

We shall also use $\mathbb{R}^n$ to denote the n-dimensional *Euclidean space*. More precisely, $\mathbb{R}^n$ is the Euclidean space with origin $= (0, 0, \ldots, 0)$, and an orthonormal basis at this point, namely (e_i), $1 \leq i \leq n$.

---

[1] Some standard references for material in this chapter are: Dieudonné, J. *Foundations of Modern Analysis*. New York: Academic Press, 1960. Edwards, C. H. *Advanced Calculus of Several Variables*. New York: Academic Press, 1973. Spivak, M. *Calculus on Manifolds*. Reading, Mass.: W. Benjamin, 1966.

The *distance* between two points $x, y \in \mathbb{R}^n$ will be denoted by $d(x, y)$ and defined by $d(x, y) := |x - y|$. Clearly $d(x, y) \geq 0$, ($d(x, y) = 0$ if and only if $x = y$) and $d(x, y) = d(y, x)$. Also, the triangle inequality for the norm implies the triangle inequality for the distance function,

$$d(x, z) \leq d(x, y) + d(y, z), \qquad x, y, z \in \mathbb{R}^n.$$

These three conditions satisfied by d imply that $\mathbb{R}^n$, with d as distance function, is a metric space.

The transformations of Euclidean space which preserve the Euclidean structure, i.e., the metric preserving transformations of $\mathbb{R}^n$, are called *isometries*. One type of isometry is a *translation*: $T_{x_0} \colon \mathbb{R}^n \to \mathbb{R}^n$ defined by $x \mapsto x + x_0$, where x_0 is a fixed element of $\mathbb{R}^n$. Another type is an *orthogonal transformation*:

$$R \colon \mathbb{R}^n \to \mathbb{R}^n, \ R \text{ is linear and } R(x) \cdot R(y) = x \cdot y, \qquad x, y \in \mathbb{R}^n.$$

If an orthogonal motion is orientation preserving (i.e., the matrix whose columns are $Re_1, \ldots, Re_n$, $i = 1, \ldots, n$, has determinant $+1$), it is a *rotation*. An example of an orthogonal motion which is not a rotation is given by the reflection

$$\rho \colon \mathbb{R}^n \to \mathbb{R}^n \qquad x \mapsto -x$$

when n is odd.

Any isometry B of Euclidean space may be written

$$B \colon \mathbb{R}^n \to \mathbb{R}^n, \qquad x \mapsto Rx + x_0$$

where $x_0 \in \mathbb{R}^n$ and R is an orthogonal motion. In other words, every isometry of Euclidean space consists of an orthogonal motion R, followed by a translation T_{x_0}. We will call R the *orthogonal component* of B. If R is a rotation we will say that B is a *congruence*. If not, we will say that B is a *symmetry*.

0.2 The Topology of Euclidean Space

The distance function d allows us, in the usual way, to define the metric topology on $\mathbb{R}^n$. For $x \in \mathbb{R}^n$ and $\epsilon > 0$, the ϵ-*ball centered at* x is denoted $B_\epsilon(x)$ and is defined by

$$B_\epsilon(x) := \{y \in \mathbb{R}^n \mid d(x, y) < \epsilon\}.$$

A set $U \subset \mathbb{R}^n$ is called *open* if for every $x \in U$ there exists an $\epsilon = \epsilon(x) > 0$ such that $B_\epsilon(x) \subset U$. A set $V \subset \mathbb{R}^n$ is *closed* if $\mathbb{R}^n \setminus V$ is open. Given a set $W \subset \mathbb{R}^n$, $\mathring{W}$ denotes its *interior*, i.e., the set of all $x \in W$ for which there exists some $\epsilon > 0$ with $B_\epsilon(x) \subset W$.

A set $U \subset \mathbb{R}^n$ is said to be a neighborhood of $x_0 \in \mathbb{R}^n$ if $x_0 \in \mathring{U}$. A mapping $F \colon U \to \mathbb{R}^n$ is *continuous at* x_0 if for every $\epsilon > 0$ there exists a $\delta > 0$ such that $F(U \cap B_\delta(x)) \subset B_\epsilon(Fx_0)$. F is said to be *continuous* if it is continuous at all $x \in U$.

Example. Linear functions are continuous

Let L be a linear function, i.e., $L(ax + by) = aL(x) + bL(y)$ for $a, b \in \mathbb{R}$, $x, y \in \mathbb{R}^n$. L may be written in terms of a matrix (a_i^j), $1 \leq i \leq n$, $1 \leq j \leq m$, where $(L(x))^j = \sum_i a_i^j x^i$. To show that L is continuous, we use the Schwarz inequality. Writing $|L|^2$ for $\sum_{i,j} (a_i^j)^2$,

$$|Lx|^2 = \sum_j \left(\sum_i a_i^j x^i\right)^2 \leq \sum_j \left(\sum_i (a_i^j)^2\right) \cdot \sum_i (x^i)^2 = |L|^2 \cdot |x|^2.$$

Therefore $|Lx - Lx_0| \leq |L| \cdot |x - x_0|$. From this, the continuity of L is easily seen. *Note*: It follows that isometries $B: \mathbb{R}^n \to \mathbb{R}^n$ are continuous: for $Bx - Bx_0 = R(x - x_0)$, R being the orthogonal component of B, and R is linear.

0.3 Differentiation in $\mathbb{R}^n$

Consider the set $L(\mathbb{R}^n, \mathbb{R}^m)$ of linear transformations from $\mathbb{R}^n$ to $\mathbb{R}^m$. This set has a natural real vector-space structure of dimension $n \cdot m$. Addition of two linear transformations L_1, L_2 is defined by adding in the range; $(L_1 + L_2)x := L_1 x + L_2 x$. Scalar multiplication by $\alpha \in \mathbb{R}$ is defined by $(\alpha L_1)x := \alpha(L_1 x)$.

In terms of the matrices (a_i^j) which represent elements $L \in L(\mathbb{R}^n, \mathbb{R}^m)$, addition corresponds to the usual matrix addition and scalar multiplication to multiplication of matrices by scalars.

The bijection of $L(\mathbb{R}^n, \mathbb{R}^m)$ onto $\mathbb{R}^{n \cdot m}$, given by considering the matrix representation (a_i^j) of a linear map L and identifying (a_i^j) with the vector $(a_1^1, \ldots, a_1^m, a_2^1, \ldots, a_2^m, \ldots, a_n^1, \ldots, a_n^m)$, is norm-preserving. The norm $|L|$ agrees with the length (= norm) of its image vector in $\mathbb{R}^{n \cdot m}$.

Let $U \subset \mathbb{R}^n$ be an open set, and suppose $F: U \to \mathbb{R}^m$ is any continuous map. F is said to be *differentiable at* $x_0 \in U$ if there exists a linear mapping $L = L(F, x_0) \in L(\mathbb{R}^n, \mathbb{R}^m)$ such that

$$\lim_{x \to x_0} \frac{|Fx - Fx_0 - L(x - x_0)|}{|x - x_0|} = 0.$$

It will be convenient to denote by $o(x)$ an arbitrary function with

$$\lim_{x \to 0} \frac{o(x)}{|x|} = 0.$$

In terms of this notation, the equation above may be rewritten as

$$|Fx - Fx_0 - L(x - x_0)| = o(x - x_0).$$

If such an $L = L(F, x_0)$ exists, it is unique. Suppose L and L' are two such linear mappings with the required properties. Then, using the triangle inequality,

$$|(L - L')(x - x_0)| = |(L - L')(x - x_0) + Fx - Fx + Fx_0 - Fx_0|$$
$$\leq |Fx - Fx_0 - L(x - x_0)| + |Fx - Fx_0 - L'(x - x_0)|$$
$$= o(x - x_0) + o(x - x_0) = o(x - x_0).$$

Thus $|(L - L')(x - x_0)|$ is $o(x - x_0)$. In particular, if $x - x_0 = re_i$, then

$$r\left(\sum_j (a_i^j - a_i'^j)^2\right)^{1/2} = o(r).$$

Therefore, $a_i^j = a_i'^j$ for all i, j.

The unique linear map $L = L(F, x_0)$ is called the *differential of F at x_0*, which will also be denoted by dF_{x_0}, or simply dF.

If A is an arbitrary (not necessarily open) set in $\mathbb{R}^n$, a mapping $F: A \to \mathbb{R}^m$ is said to be differentiable on A if there exists an open set $U \subset \mathbb{R}^n$ containing A and a mapping $G: U \to \mathbb{R}^n$ such that $G|_A = F$, and G is differentiable at each $x_0 \in U$.

Examples of differentiable mappings

1. $L: \mathbb{R}^n \to \mathbb{R}^m$, any linear map. $dL_x = L$, for all $x \in \mathbb{R}^n$.
2. $B: \mathbb{R}^n \to \mathbb{R}^m$, an isometry. $dB_x = R$, the orthogonal component of B.
3. All the elementary functions encountered in calculus of one variable are differentiable; polynomials, rational functions, trigonometric functions, the exponential and logarithm.
4. The maps $(x, y) \mapsto x \cdot y$ from $\mathbb{R}^n \times \mathbb{R}^n$ into $\mathbb{R}$ and $x \mapsto |x|^2$ from $\mathbb{R}^n$ into $\mathbb{R}$ are differentiable.
5. The familiar vector cross-product $(x, y) \mapsto x \times y \in \mathbb{R}^3$, considered as a map from $\mathbb{R}^3 \times \mathbb{R}^3$ into $\mathbb{R}^3$, is differentiable. In terms of a basis for $\mathbb{R}^3$, if $x = (x_1, x_2, x_3)$ and $y = (y_1, y_2, y_3)$, then $x \times y = (x_2 y_3 - x_3 y_2, x_3 y_1 - x_1 y_3, x_1 y_2 - x_2 y_1)$.

It is an easy exercise to prove that the composition of two differentiable mappings is differentiable.

A mapping $F: U \to \mathbb{R}^m$, U open in $\mathbb{R}^n$, is said to be *continuously differentiable*, or C^1, if F is differentiable at each $x \in U$ and the map $dF: U \to L(\mathbb{R}^n, \mathbb{R}^m)$, given by $x \to dF_x$, is continuous.

A mapping $F: U \to \mathbb{R}^m$, $U \subset \mathbb{R}^n$ is said to be *twice continuously differentiable*, or C^2, if $dF: U \to L(\mathbb{R}^n, \mathbb{R}^m)$ is differentiable, and its derivative is continuous.

In an analogous manner, we may define k-times continuously differentiable mappings, or C^k mappings. If f is k-times differentiable for any $k = 1, 2, \ldots$, f is said to be C^∞ (read "C infinity"). Sometimes we will refer to C^∞ mappings as differentiable mappings when there is no possibility of confusion.

If $U \subset \mathbb{R}^m$, $V \subset \mathbb{R}^n$ are open sets and $F: U \to V$ is a bijective, differentiable function such that $F^{-1}: V \to U$ is also differentiable, then F is called a *diffeomorphism* (between U and V).

If $F: U \to \mathbb{R}^m$, $U \subset \mathbb{R}^n$ is differentiable, then the m coordinate functions $F^j(x^1, \ldots, x^n)$ have partial derivatives $\partial F^j/\partial x^i = F^j_{x^i}$ with respect to each of the n coordinates x^i. From our definition of $dF_{x_0}: \mathbb{R}^m \to \mathbb{R}^n$, it follows that the matrix of this linear map is given by the matrix of first derivatives of F at x_0, $(F^j_{x^i})_{x_0}$, the familiar Jacobian matrix.

The differential $d^2F = d(dF)$ of the differentiable function $dF: U \to L(\mathbb{R}^n, \mathbb{R}^m)$ at the point $x_0 \in U$ has the following matrix representation: dF is determined by the $n \cdot m$ real valued functions $\partial F^j/\partial x^i$. Therefore $d^2F_{x_0}$ is determined by the $(m \times n \cdot m)$-matrix $(\partial^2 F^j/\partial x^i \partial x^k)|_{x_0}$. The row-index in this notation is $\{^j_i\}$ and k is the column-index. (The pairs $\{^j_i\}$ are ordered lexicographically.)

0.4 Tangent Space

The concept of a tangent space will play a fundamental role in our study of differential geometry. For $x_0 \in \mathbb{R}^n$, the *tangent space of* $\mathbb{R}^n$ *at* x_0, written $T_{x_0}\mathbb{R}^n$ or $\mathbb{R}^n_{x_0}$, is the n-dimensional vector-space whose elements consist of pairs $(x_0, x) \in \{x_0\} \times \mathbb{R}^n$. The vector-space structure is defined by means of the bijection

$$T_{x_0}\mathbb{R}^n \to \mathbb{R}^n, \qquad (x_0, x) \mapsto x,$$

i.e., $(x_0, x) + (x_0, y) = (x_0, x + y)$ and $a(x_0, x) = (x_0, ax)$.

Let U be a subset of $\mathbb{R}^n$. *The tangent bundle of U*, denoted TU, is the disjoint union of the tangent spaces $T_{x_0}\mathbb{R}^n$, $x_0 \in U$, together with the canonical projection $\pi: TU \to U$, given by $(x_0, x) \mapsto x_0$. TU is in 1-1 correspondence with $U \times \mathbb{R}^n$ via the bijection

$$(x_0, x) \in T_{x_0}\mathbb{R}^n \subset TU \mapsto (x_0, x) \in U \times \mathbb{R}^n.$$

In view of the generalizations we will make in subsequent chapters, the interpretation of TU as the disjoint union of the tangent spaces $T_{x_0}\mathbb{R}^n$, $x_0 \in U$, is preferable to that of TU as $U \times \mathbb{R}^n$. On the other hand, the interpretation of TU as $U \times \mathbb{R}^n$ shows that TU may be considered as a subset of $\mathbb{R}^n \times \mathbb{R}^n = \mathbb{R}^{2n}$. If U is open, then $U \times \mathbb{R}^n$ is also open in $\mathbb{R}^{2n}$, so it is clear what it means for a function $G: TU \to \mathbb{R}^k$ to be continuous or differentiable. We may now define the notion of the *differential* of a differentiable mapping $F: U \to \mathbb{R}^m$ in terms of the tangent bundle.

Let U be an open set in $\mathbb{R}^n$ and let $F: U \to \mathbb{R}^m$ be a differentiable function. For each $x_0 \in U$ we define the map $TF_{x_0}: T_{x_0}\mathbb{R}^n \to T_{F(x_0)}\mathbb{R}^m$ by $(x_0, x) \mapsto (F(x_0), dF_{x_0}(x))$. The map $TF: TU \to T\mathbb{R}^m$ is now defined by $TF|T_{x_0}\mathbb{R}^n := TF_{x_0}$. TF is called the *differential* of F.

A word about notation: If we identify $T_{x_0}\mathbb{R}^n$ with $\mathbb{R}^n$ in the canonical way, and likewise $T_{F(x_0)}\mathbb{R}^m$ with $\mathbb{R}^m$, then instead of $TF_{x_0}: T_{x_0}\mathbb{R}^n \to T_{F(x_0)}\mathbb{R}^m$ we write $dF_{x_0}: \mathbb{R}^n \to \mathbb{R}^m$.

0.5 Local Behavior of Differentiable Functions (Injective and Surjective Functions)

We shall need to use the following basic theorem:

0.5.1 Theorem (Inverse function theorem). *Let U be an open neighborhood of $0 \in \mathbb{R}^n$. Suppose $F: U \to \mathbb{R}^n$ is a differentiable function with $F(0) = 0 \in \mathbb{R}^n$. If $dF_0: \mathbb{R}^n \to \mathbb{R}^n$ is bijective, then there is an open neighborhood $U' \subset U$ of 0 such that $F|_{U'}: U' \to FU'$ is a diffeomorphism.*

Such a function F is said to be a *local diffeomorphism* (or, more precisely, a local diffeomorphism at 0).

In order to state and prove an important consequence of the inverse function theorem, it is necessary to recall some facts about linear maps. A linear map $L: \mathbb{R}^n \to \mathbb{R}^m$ is injective, or 1-1, if and only if $\ker L := \{x \in \mathbb{R}^n \mid Lx = 0\} = \{0\}$. This is equivalent, in turn, to the requirement that $\mathbb{R}^m$ has a direct sum decomposition $\mathbb{R}^m = \mathbb{R}'^n \oplus \mathbb{R}''^{m-n}$ (into subspaces of dimension n and $m - n$, respectively) such that $L: \mathbb{R}^n \to \mathbb{R}'^n$ is a bijection.

Similarly, a linear map $L: \mathbb{R}^n \to \mathbb{R}^m$ is surjective, or onto, if and only if $n - m = \dim \ker L$. This condition is equivalent to the existence of a direct sum decomposition $\mathbb{R}^n = \mathbb{R}'^m \oplus \mathbb{R}''^{n-m}$ into subspaces of dimension m and $n - m$, respectively, such that $\mathbb{R}''^{n-m} = \ker L$ and $L|_{\mathbb{R}'^m}: \mathbb{R}'^m \to \mathbb{R}^m$ is a bijection.

The next theorem shows that, locally, differentiable functions behave in a manner analogous to linear maps, at least with respect to the injectivity and surjectivity properties described above.

0.5.2 Theorem (Local linearization of differentiable mappings). *Let U be an open neighborhood of $0 \in \mathbb{R}^n$. Suppose $F: U \to \mathbb{R}^m$ is a differentiable function with $F(0) = 0$.*
  i) *If $TF_0: T_0 \mathbb{R}^n \to T_0 \mathbb{R}^m$ is injective, then there exists a diffeomorphism g of a neighborhood W of $0 \in \mathbb{R}^m$ onto a neighborhood $g(W)$ of $0 \in \mathbb{R}^m$ such that $g \circ F$ is an injective linear map from some neighborhood of $0 \in \mathbb{R}^n$ into $\mathbb{R}^m$. In fact, $g \circ F(x_1, \ldots, x_n) = (x_1, \ldots, x_n, 0, \ldots, 0)$.*
  ii) *If $TF_0: T_0 \mathbb{R}^n \to T_0 \mathbb{R}^m$ is surjective, there exists a diffeomorphism h of a neighborhood V of $0 \in \mathbb{R}^n$ onto a neighborhood $h(V)$ of $0 \in \mathbb{R}^n$ such that $F \circ h$ is a surjective linear map from some neighborhood of $0 \in \mathbb{R}^n$ onto a neighborhood of $0 \in \mathbb{R}^m$. In fact, $F \circ h(x_1, \ldots, x_m, \ldots, x_n) = (x_1, \ldots, x_m)$.*

Remark. The converse of each of the above statements is clearly true.

PROOF. i) Suppose $dF_0: \mathbb{R}^n \to \mathbb{R}^m$ is injective. Write $\mathbb{R}^m = \mathbb{R}'^n \oplus \mathbb{R}''^{m-n}$ with $dF_0(\mathbb{R}^n) = \mathbb{R}'^n$. Define $\tilde{g}: \mathbb{R}^m = \mathbb{R}'^n \oplus \mathbb{R}''^{m-n} \to \mathbb{R}^m = \mathbb{R}'^n \oplus \mathbb{R}''^{m-n}$ in a neighborhood of 0 by $v = (v', v'') \mapsto F(v') + (0, v'')$. Here the $\mathbb{R}'^n$ on the left-hand side is identified with $\mathbb{R}^n$. Clearly, $d\tilde{g}_0 = dF_0 + \text{id} \mid \mathbb{R}''^{m-n}$.

Therefore $d\tilde{g}_0$ is bijective and we may use the inverse function theorem (0.5.1) to assert the existence of a local differentiable inverse $g = \tilde{g}^{-1}$.

Since $g \circ \tilde{g} = \text{id}$, $g \circ \tilde{g} \mid \mathbb{R}'^n = \text{id} \mid \mathbb{R}'^n$ locally, and thus $g \circ F(v') = (v', 0)$. This proves $g \circ F$ is a linear injective function from a neighborhood of 0 in $\mathbb{R}^n$ into $\mathbb{R}'^n \subset \mathbb{R}'^n \oplus \mathbb{R}''^{m-n} = \mathbb{R}^m$.

ii) Suppose $dF_0 : \mathbb{R}^n \to \mathbb{R}^m$ is surjective. Decomposing $\mathbb{R}^n = \mathbb{R}'^m \oplus \mathbb{R}''^{n-m}$ so that $dF_0 \mid \mathbb{R}'^m : \mathbb{R}'^m \to \mathbb{R}^m$ is a bijection, define $\tilde{h}: \mathbb{R}^n = \mathbb{R}'^m \oplus \mathbb{R}''^{n-m} \to \mathbb{R}^n = \mathbb{R}'^m \oplus \mathbb{R}''^{n-m}$ in a neighborhood of zero by $v = (v', v'') \mapsto (Fv, v'')$. Here we have identified $\mathbb{R}'^m$ on the right-hand side with $\mathbb{R}^m$.

Since $d\tilde{h}_0 = dF_0 \mid \mathbb{R}'^m + \text{id} \mid \mathbb{R}''^{n-m}$ is bijective, $\tilde{h}$ has a local inverse $h = \tilde{h}^{-1}$. Since $h \circ \tilde{h} = \text{id}$ locally, $h(F(v', v''), v'') = (v', v'')$ and therefore $F \circ h(F(v', v''), v'') = F(v', v'')$. This means that $F \circ h$ is given locally by the projection $\mathbb{R}^n = \mathbb{R}'^m \oplus \mathbb{R}''^{n-m} \to \mathbb{R}'^m$ onto the first m coordinates, which, of course, is linear and surjective. □

0.6 Exercise

Prove that any distance-preserving mapping $B: \mathbb{R}^n \to \mathbb{R}^n$ may be written in the form

$$Bx = Rx + x_0,$$

an orthogonal motion followed by a translation.

1 Curves

1.1 Definitions

1.1.1 Definitions. Let $I \subseteq \mathbb{R}$ be an interval. For our purposes, a (parametrized) *curve in* $\mathbb{R}^n$ is a C^∞ mapping $c: I \to \mathbb{R}^n$. c will be said to be *regular* if for all $t \in I$, $\dot{c}(t) \neq 0$.

Remarks. 1. If I is not an open interval, we need to make explicit what it means for c to be C^∞. There exists an open interval I^* containing I and a C^∞ mapping $c^*: I^* \to \mathbb{R}^n$ such that $c = c^*|I$.

2. The variable $t \in I$ is called the *parameter* of the curve.

3. The tangent space $\mathbb{R}_{t_0} = T_{t_0}\mathbb{R}$ of $\mathbb{R}$ at $t_0 \in I$ has a distinguished basis $1 = (t_0, 1)$. As an alternate notation we will sometimes write d/dt for $(t_0, 1) = 1$.

4. If $c: I \to \mathbb{R}^n$ is a curve, the vector $dc_{t_0}(1) \in T_{c(t_0)}\mathbb{R}^n$ is well defined. Since $|c(t) - c(t_0) - dc_{t_0}(1)(t - t_0)| = o(t - t_0)$, it follows immediately that $dc_{t_0}(1) = \lim_{t \to t_0}[c(t) - c(t_0)]/(t - t_0) = \dot{c}(t_0)$, the derivative of the $\mathbb{R}^n$-valued function $c(t)$ at $t_0 \in I$.

1.1.2 Definitions. i) A *vector field along* $c: I \to \mathbb{R}^n$ is a differentiable mapping $X: I \to \mathbb{R}^n$. The vector $X(t)$, that is the value of X at a given $t \in I$, will be thought of as lying in the copy of $\mathbb{R}^n$ identified with $T_{c(t)}\mathbb{R}^n$ (see Figure 1.1).

ii) The *tangent vector field of* $c: I \to \mathbb{R}^n$ is the vector field along $c: I \to \mathbb{R}^n$ given by $t \mapsto \dot{c}(t)$.

1.1.3 Definition. Let $c: I \to \mathbb{R}^n$, $\tilde{c}: \tilde{I} \to \mathbb{R}^n$ be two curves. A diffeomorphism $\phi: \tilde{I} \to I$ such that $\tilde{c} = c \circ \phi$ is called a *parameter transformation* or a *change of variables* relating c to $\tilde{c}$. The map ϕ is called *orientation preserving* if $\phi' > 0$.

1.1 Definitions

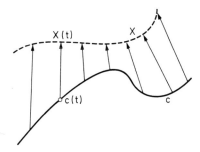

Figure 1.1

Remark. Relationship by a parameter transformation is clearly an equivalence relation on the set of all curves in $\mathbb{R}^n$. An equivalence class of curves is called an *unparameterized curve*.

1.1.4 Definitions. i) The curve $c(t)$, $t \in I$, is said to be *parameterized by arc length* if $|\dot{c}(t)| = 1$. We will sometimes refer to such a curve as a *unit-speed* curve.

ii) *The length of c is given by the integral* $L(c) := \int_I |\dot{c}(t)| \, dt$.

iii) The integral $E(c) := \frac{1}{2} \int_I \dot{c}(t)^2 \, dt$ is called the *energy integral of c* or, simply, *the energy of c*.

1.1.5 Proposition. *Every regular curve $c: I \to \mathbb{R}^n$ can be parameterized by arc length. In other words, given a regular curve $c: I \to \mathbb{R}^n$ there is a change of variables $\phi: J \to I$ such that $|(c \circ \phi)'(s)| = 1$.*

PROOF. The desired equation for ϕ is $|dc/ds| = |dc/dt| \cdot |d\phi/ds| = 1$. Define $s(t) = \int_{t_0}^{t} |\dot{c}(t')| \, dt'$, $t_0 \in I$, and let $s(t) = \phi^{-1}(t)$. Since c is regular, ϕ exists

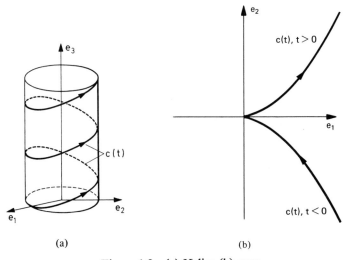

Figure 1.2 (a) Helix; (b) cusp

1 Curves

and satisfies the desired equation. Clearly, $c \circ \phi$ is parameterized by arc length. □

Examples

1. *Straight line.* For $v, v_0 \in \mathbb{R}^n$ let $c(t) = tv + v_0$, $t \in \mathbb{R}$. The curve $c(t)$ is regular if and only if $v \neq 0$ and, in this case, is a straight line.
2. *Circle and helix.* $c(t) = (a \cos t, a \sin t, bt)$, $a, b, t \in \mathbb{R}$, $a^2 + b^2 \neq 0$. When $b = 0$, $c(t)$ is a plane circle of radius a. When $a = 0$, $c(t)$ is a straight line. In general, $c(t)$ is a helix. In all cases, $c(t)$ is a regular curve.
3. *Parameterization of a cusp.* The curve $c(t) = (t^2, t^3)$, $t \in \mathbb{R}$, is regular when $t \neq 0$. The image of $c(t)$ is a cusp.
4. *Another parameterization of a straight line.* The curve $c(t) = (t^3, t^3)$, $t \in \mathbb{R}$, is regular when $t \neq 0$. The image of $c(t)$ is a straight line.

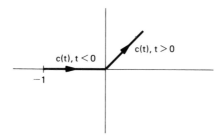

Figure 1.3 Image of c

1.2 The Frenet Frame

1.2.1 Definition. Let $c: I \to \mathbb{R}^n$ be a curve. i) A *moving n-frame* along c is a collection of n differentiable mappings

$$e_i: I \to \mathbb{R}^n, \quad 1 \leq i \leq n,$$

such that for all $t \in I$, $e_i(t) \cdot e_j(t) = \delta_{ij}$, where $\delta_{ij} = \{\begin{smallmatrix}1; i=j\\0; i\neq j\end{smallmatrix}\}$. Each $e_i(t)$ is a vector field along c, and $e_i(t)$ is considered as a vector in $T_{c(t)}\mathbb{R}^n$.

ii) A moving n-frame is called a *Frenet-n-frame*, or simply *Frenet frame*, if for all k, $1 \leq k \leq n$, the kth derivative $c^{(k)}(t)$ of $c(t)$ lies in the span of the vectors $e_1(t), \ldots, e_k(t)$.

Remark. Not every curve possesses a Frenet-n-frame. Consider

$$c: \mathbb{R} \to \mathbb{R}^2, \quad c(t) = \begin{cases} (-e^{-1/t^2}, 0), & \text{if } t < 0 \\ (e^{-1/t^2}, e^{-1/t^2}), & \text{if } t > 0 \\ (0, 0), & \text{if } t = 0 \end{cases}.$$

Because the image of c has a crease at $(0, 0)$ it is impossible to find a differentiable unit vector field $e_1(t)$ along c such that $\dot{c}(t) = |\dot{c}(t)|e_1(t)$.

1.2.2 Proposition (The existence and uniqueness of a distinguished Frenet-frame). *Let $c: I \in \mathbb{R}^n$ be a curve such that for all $t \in I$, the vectors $\dot{c}(t), c^{(2)}(t), \ldots, c^{(n-1)}(t)$ are linearly independent. Then there exists a unique Frenet-frame with the following properties:*

i) *For $1 \leq k \leq n - 1$, $\dot{c}(t), \ldots, c^{(k)}(t)$ and $e_1(t), \ldots, e_k(t)$ have the same orientation.*

ii) *$e_1(t), \ldots, e_n(t)$ has the positive orientation.*

This frame is called the *distinguished Frenet-frame*.

Remark. Recall that two bases for a real vector space have the same orientation provided the linear transformation taking one basis into the other has positive determinant. A basis for $\mathbb{R}^n$ is *positively oriented* if it has the same orientation as the canonical basis of $\mathbb{R}^n$.

PROOF. We will use the Gram-Schmidt orthogonalization process. The assumption that $\dot{c}(t), \ddot{c}(t), \ldots$ are linearly independent implies that $\dot{c}(t) \neq 0$ and so we may set $e_1(t) = \dot{c}(t)/|\dot{c}(t)|$. Suppose $e_1(t), \ldots, e_{j-1}(t), j < n$, are defined. Let $\tilde{e}_j(t)$ be defined by

$$\tilde{e}_j(t) := -\sum_{k=1}^{j-1} (c^{(j)}(t) \cdot e_k(t)) e_k(t) + c^{(j)}(t)$$

and let $e_j(t) := \tilde{e}_j(t)/|\tilde{e}_j(t)|$.

Clearly, the $e_j(t)$, $j < n$, are well defined and satisfy the first assertion of the theorem. Furthermore, we may define $e_n(t)$ so that $e_1(t), \ldots, e_n(t)$ has positive orientation. The differentiability of $e_j(t)$, $j < n$, is clear from its definition. To see that $e_n(t)$ is differentiable, observe that each of the components $e_n^i(t)$, $1 \leq i \leq n$, of $e_n(t)$ may be expressed as the determinant of a minor of rank $(n-1)$ in the $n \times (n-1)$-matrix $(e_j^i(t))$, $1 \leq i \leq n$, $1 \leq j \leq n-1$. □

1.3 The Frenet Equations

1.3.1 Proposition. *Let $c(t)$, $t \in I$, be a curve in $\mathbb{R}^n$ together with a moving frame $(e_i(t))$, $1 \leq i \leq n$, $t \in I$. Then the following equations for the derivatives hold:*

$$\dot{c}(t) = \sum_i \alpha_i(t) e_i(t),$$

$$\dot{e}_i(t) = \sum_j \omega_{ij}(t) e_j(t),$$

where

(*) $$\omega_{ij}(t) := \dot{e}_i(t) \cdot e_j(t) = -\omega_{ji}(t).$$

If $(e_i(t))$ is the distinguished Frenet-frame defined in (1.2.2),

(**) $$\alpha_1(t) = |\dot{c}(t)|, \quad \alpha_i(t) = 0 \quad \text{for } i > 1,$$

and

$$\omega_{ij}(t) = 0 \quad \text{for } j > i + 1.$$

1 Curves

PROOF. Equation (*) follows from differentiating $e_i(t) \cdot e_j(t) = \delta_{ij}$.

Equations (**) hold for distinguished Frenet-frames because the condition that $e_i(t)$ is a linear combination of $\dot{c}(t), \ldots, c^{(i)}(t)$ implies that $\dot{e}_i(t)$ is a linear combination of $\ddot{c}(t), \ldots, c^{(i+1)}(t)$ and hence of $e_1(t), \ldots, e_{i+1}(t)$. □

Remark. If $\omega(t)$ denotes the one-parameter family of matrices $(\omega_{ij}(t))$, $1 \le i, j \le n$, we may write the n equations

$$\dot{e}_i(t) = \sum_j \omega_{ij}(t) e_j(t)$$

as

$$\dot{e}(t) = \omega(t) e(t),$$

where $e(t)$ is the matrix whose rows are the vectors $e_i(t)$. Equation (*) then says: ω is skew-symmetric. If, in addition, $(e_i(t))$ is a distinguished Frenet-frame, (**) implies that ω is of the form

$$\omega = \begin{pmatrix} 0 & \omega_{12} & 0 & & & \cdots & & 0 \\ -\omega_{12} & 0 & \omega_{23} & 0 & & & & 0 \\ 0 & -\omega_{23} & 0 & \omega_{34} & 0 & \cdots & & 0 \\ \vdots & & & & \ddots & & & \vdots \\ 0 & & \cdots & & & 0 & & \omega_{n-1,n} \\ 0 & & \cdots & & 0 & -\omega_{n-1,n} & & 0 \end{pmatrix}.$$

The next proposition proves that these differential equations are invariant under isometries of $\mathbb{R}^n$, and establishes how these equations transform under a change of variables.

1.3.2 Proposition. i) *Let $c: I \to \mathbb{R}^n$ be a curve and $B: \mathbb{R}^n \to \mathbb{R}^n$ an isometry of $\mathbb{R}^n$ whose orthogonal component is R. Let $\tilde{c} = B \circ c: I \to \mathbb{R}^n$, and let $(e_i(t))$, $i = 1, \ldots, n$, be a moving frame on c. Then $(\tilde{e}_i(t)) := (Re_i(t))$, $i = 1, \ldots, n$, is a moving frame on $\tilde{c}$ and if $\tilde{\omega}_{ij}(t)$ are the coefficients of the associated Frenet equation for $\tilde{c}$, $(\tilde{e}_i(t))$, then*

$$|\dot{\tilde{c}}(t)| = |\dot{c}(t)|$$

and

$$\tilde{\omega}_{ij}(t) = \omega_{ij}(t).$$

ii) *Let $c: I \to \mathbb{R}^n$ and $\tilde{c}: J \to \mathbb{R}^n$ be curves in $\mathbb{R}^n$, related by the orientation-preserving change of variables ϕ. In other words,*

$$\tilde{c} = c \circ \phi, \qquad \phi'(s) > 0.$$

Let $(e_i(t))$, $i = 1, \ldots, n$, be a moving frame on c. Then $(\tilde{e}_i(s)) = (e_i \circ \phi(s))$, $i = 1, \ldots, n$, is a moving frame on $\tilde{c}$. If $|\tilde{c}'(s)| \ne 0$, then

$$\frac{\tilde{\omega}_{ij}(s)}{|\tilde{c}'(s)|} = \frac{\omega_{ij}(\phi(s))}{|\dot{c}(\phi(s))|}.$$

PROOF. i) $\tilde{\omega}_{ij}(t) = \dot{\tilde{e}}_i(t) \cdot \tilde{e}_j(t) = R\dot{e}_i(t) \cdot Re_j(t) = \dot{e}_i(t) \cdot e_j(t) = \omega_{ij}(t)$.

ii) $\dfrac{\tilde{\omega}_{ij}(s)}{|\tilde{c}'(s)|} = \tilde{e}'_i(s) \cdot \dfrac{\tilde{e}_j(s)}{|\tilde{c}'(s)|} = \dot{e}_i(\phi(s))\phi'(s) \cdot \dfrac{e_j(\phi(s))}{|\dot{c}(\phi(s))|\phi'(s)} = \dfrac{\omega_{ij}(\phi(s))}{|\dot{c}(\phi(s))|}$. □

1.3.3 Definition. Let $c: I \to \mathbb{R}^n$ be a curve satisfying the conditions of (1.2.2), and consider its distinguished Frenet-frame. The *i*th *curvature of c*, $i = 1, 2, \ldots, n-1$, is the function

$$\kappa_i(t) := \frac{\omega_{i,i+1}(t)}{|\dot{c}(t)|}.$$

Note that for the distinguished Frenet-frame we may now write the matrix ω as

$$\omega = |\dot{c}| \begin{pmatrix} 0 & \kappa_1 & 0 & \cdots & 0 \\ -\kappa_1 & 0 & \kappa_2 & \cdots & 0 \\ \vdots & -\kappa_2 & \ddots & \ddots & \\ 0 & \cdots & -\kappa_{n-2} & 0 & \kappa_{n-1} \\ 0 & \cdots & & -\kappa_{n-1} & 0 \end{pmatrix}.$$

Let us establish a simple fact about the curvature functions, κ_i, $i < n - 1$. Namely: they are positive. Remember, we have only defined the κ_i for curves satisfying the nondegeneracy conditions of (1.2.2).

1.3.4 Proposition. *Let $\kappa_i(t)$, $1 \leq i \leq n - 1$, be the curvature functions defined in (1.3.3). Then $\kappa_i(t) > 0$ for $1 \leq i \leq n - 2$.*

PROOF. By construction (in (1.2.2)),

$$c^{(k)} = \sum_{l=1}^{k} a_{kl} e_l \quad \text{and} \quad e_k = \sum_{l=1}^{k} b_{kl} c^{(l)} \quad \text{with} \quad a_{kk} > 0$$

(and so $b_{kk} = a_{kk}^{-1} > 0$) for $1 \leq k \leq n-1$. Therefore for $1 \leq i \leq n-2$,

$$|\dot{c}|\kappa_i = \omega_{i,i+1} = \dot{e}_i \cdot e_{i+1} = b_{ii} c^{(i+1)} \cdot e_{i+1} = b_{ii} a_{i+1,i+1} > 0. \quad \square$$

We now explore to what extent these curvature functions determine curves satisfying the nondegeneracy conditions of (1.2.2).

1.3.5 Theorem. *Let $c: I \to \mathbb{R}^n$ and $\tilde{c}: I \to \mathbb{R}^n$ be two curves satisfying the hypotheses of (1.2.2), insuring the existence of a unique distinguished Frenet-frame. Denote these Frenet-frames by $(e_i(t))$ and $(\tilde{e}_i(t))$, respectively, $1 \leq i \leq n$. Suppose, relative to these frames, that $\kappa_i(t) = \tilde{\kappa}_i(t)$, $1 \leq i \leq n - 1$, and assume $|\dot{c}(t)| = |\dot{\tilde{c}}(t)|$. Then there exists a unique isometry $B: \mathbb{R}^n \to \mathbb{R}^n$ such that*

$$\tilde{c} = B \circ c.$$

1 Curves

Furthermore, B is a congruence; its orthogonal component has determinant +1 (a rotation).

PROOF. Fix $t_0 \in I$. There is precisely one isometry B satisfying

$$Bc(t_0) = \tilde{c}(t_0),$$
$$Re_i(t_0) = \tilde{e}_i(t_0), \quad 1 \le i \le n,$$

where R is the orthogonal component of B. Since both Frenet-frames are positively oriented, R has determinant equal to $+1$.

From the hypotheses we have $\tilde{\omega}_{ij}(t) = \omega_{ij}(t)$, which implies

$$\dot{\tilde{e}}_i(t) = \sum_j \omega_{ij}(t)\tilde{e}_j(t).$$

On the other hand,

$$R\dot{e}_i(t) = \sum_j \omega_{ij}(t)Re_j(t).$$

We see that $\tilde{e}_i(t)$ and $Re_i(t)$ satisfy the same system of linear differential equations. Since they are equal at $t = t_0$, $Re_i(t) = \tilde{e}_i(t)$ for all $t \in I$. In particular, $R\dot{c}(t) = |\dot{c}(t)|Re_1(t) = |\dot{\tilde{c}}(t)|\tilde{e}_1(t) = \dot{\tilde{c}}(t)$. Thus

$$Bc(t) - Bc(t_0) = \int_{t_0}^{t} R\dot{c}(t)\, dt = \int_{t_0}^{t} \dot{\tilde{c}}(t)\, dt = \tilde{c}(t) - \tilde{c}(t_0),$$

which proves $Bc(t) = \tilde{c}(t)$.

To see that B is unique, let B' be another isometry satisfying $B' \circ c = \tilde{c}$. Then B' must transform the distinguished Frenet-frame of c into that of $\tilde{c}$. In addition, $B' \circ c(t_0) = \tilde{c}(t_0)$, so B and B' have the same translation component and the same orthogonal component. Therefore $B = B'$. □

1.3.6 Theorem (Existence of curves with prescribed curvature functions). *Let $\kappa_1(s), \ldots, \kappa_{n-1}(s)$ be differentiable functions defined on a neighborhood $0 \in \mathbb{R}$ with $\kappa_i(s) > 0$, $1 \le i \le n - 2$. Then there exists an interval I containing 0 and a unit speed curve $c: I \to \mathbb{R}^n$ which satisfies the conditions of (1.2.2) and whose ith curvature function is $\kappa_i(s)$, $1 \le i \le n - 1$.*

PROOF. Consider the matrix-valued function

$$A(s) = \begin{pmatrix} 0 & \kappa_1(s) & & \cdots & & 0 \\ -\kappa_1(s) & 0 & & & & \\ & & \ddots & & & \\ & & & & 0 & \kappa_{n-1}(s) \\ 0 & & \cdots & & -\kappa_{n-1}(s) & 0 \end{pmatrix}$$

and the linear system of differential equations $X'(s) = A(s) \cdot X(s)$, $X(0) = \text{Id}$, where $X(s)$ is an $n \times n$ matrix-valued function, Id is the $n \times n$ identity matrix and the multiplication is matrix multiplication. By standard results in differential equations (e.g., Hurewicz, W. *Lectures on ordinary differential equations.* MIT Press, Cambridge, Mass. (1958) p. 28), there exists a solution $X(s)$ defined on some interval I containing $0 \in \mathbb{R}$.

Since $A(s)$ is skew-symmetric $({}^tA(s) = -A(s))$, $({}^tX(s) \cdot X(s))' = {}^t(A(s) \cdot X(s)) \cdot X(s) + {}^tX(s) \cdot A(s) \cdot X(s) = {}^tX(s) \cdot {}^tA(s) \cdot X(s) + {}^tX(s) \cdot A(s) \cdot X(s) = 0$. Thus ${}^tX(s) \cdot X(s)$ is a constant matrix and must be equal to its value at $s = 0$, namely the identity matrix. Therefore $X(s)$ is an orthogonal matrix. Let $T(s)$ be the first column of $X(s)$ and define

$$c(s) = \int_0^s T(\tau)d\tau, \quad s \in I,$$

the integration being done component-wise. One can now check directly that $c(s)$ is a unit speed curve with distinguished Frenet frame $X(s)$ and curvature functions $\kappa_i(s)$, $1 \le i \le n - 1$. $\square$

1.4 Plane Curves; Local Theory

In this section we will investigate plane curves; $c: I \to \mathbb{R}^2$. We will assume throughout that $\dot{c}(t) \ne 0$, i.e., c is regular. For plane curves this hypothesis is equivalent to (1.2.2). Thus we may always construct the distinguished Frenet-frame, and we shall always choose this frame as the moving 2-frame on our curve c.

The Frenet equations of (1.3.1) for a plane curve are

$$\dot{c}(t) = |\dot{c}(t)|e_1(t)$$
$$\dot{e}_1(t) = \omega_{12}(t)e_2(t)$$
$$\dot{e}_2(t) = -\omega_{12}(t)e_1(t),$$

or

$$\dot{c}(t) = |\dot{c}(t)|e_1(t)$$
$$\dot{e}(t) = \begin{pmatrix} 0 & \omega_{12}(t) \\ -\omega_{12}(t) & 0 \end{pmatrix} e(t),$$

and there is only one curvature:

$$\kappa(t) := \frac{\omega_{12}(t)}{|\dot{c}(t)|}.$$

In the special case that $|\dot{c}(t)| = 1$, $\dot{c}(t) = e_1(t)$ and

$$\ddot{c}(t) = \dot{e}_1(t) = \omega_{12}(t)e_2(t) = \kappa(t)e_2(t),$$

so $|\kappa(t)| = |\ddot{c}(t)|$.

The sign of $\kappa(t)$ is positive (negative) when $e_2(t)$ and $\ddot{c}(t)$ make an acute (obtuse) angle with each other.

Expressed graphically: $\kappa(t) > 0$ ($\kappa(t) < 0$) means that $e_2(t)$ points toward the convex (concave) side of the curve c at $c(t)$.

1 Curves

Example. Graph of the sine
$$c(t) = (t, \sin t), \quad \text{for } t \in \mathbb{R},$$
$$\kappa(t) < 0 \quad \text{for } t \in (0, \pi),$$
$$\kappa(t) > 0 \quad \text{for } t \in (\pi, 2\pi).$$

It is possible that $\kappa(t) = 0$. If, in addition, $\dot{\kappa}(t) \neq 0$ (and hence the zero of κ is isolated) $c(t)$ is called an *inflection point* of the curve. In the example above, $c(0)$ and $c(\pi)$ are inflection points.

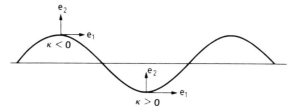

Figure 1.4   The sine curve

The curvature function for plane curves has the following geometric interpretation: Fix some vector v of unit length. Define $\theta(t)$ by
$$\cos \theta(t) = e_1(t) \cdot v,$$
$$\sin \theta(t) = -e_2(t) \cdot v.$$

Thus $\theta(t)$ is, up to a multiple of 2π, the angle from v to $e_1(t)$ measured in the positive direction. In a sufficiently small neighborhood of any parameter value $t_0 \in I$, $\theta(t)$ may be defined so that it is continuous. Doing this will also make $\theta(t)$ differentiable in that neighborhood. Clearly, $\dot{\theta}(t)$ is a well-defined function, independent of the choices involved in defining $\theta(t)$.

1.4.1 Proposition. *Suppose $\theta(t)$ is locally defined as above. Then*
$$\dot{\theta}(t) = \omega_{12}(t) = \kappa(t)|\dot{c}(t)|.$$
In the case that $|\dot{c}(t)| = 1$, $\kappa(t) = \dot{\theta}(t)$.

PROOF. The proposition is an immediate consequence of differentiating the defining equations for $\theta(t)$:
$$-\sin \theta(t)\dot{\theta}(t) = \omega_{12}(t)e_2(t) \cdot v = -\sin \theta(t)\omega_{12}(t),$$
$$\cos \theta(t)\dot{\theta}(t) = \omega_{12}(t)e_1(t) \cdot v = \cos \theta(t)\omega_{12}(t). \quad \square$$

1.4.2 Proposition (Characterization of straight lines). *For plane curves, the following conditions are equivalent.*
  i) $\kappa(t) = 0$ *for all $t \in I$.*
  ii) *There exists a parameterization of c of the form*
$$c(t) = (t - t_0)v + v_0, \quad \text{where} \quad t_0 \in \mathbb{R}, v, v_0 \in \mathbb{R}^2, v \neq 0,$$
i.e., a straight line.

PROOF. We may assume $|\dot{c}(t)| = 1$. If $\kappa(t) = 0$ then $\ddot{c}(t) = 0$. Therefore $c(t) = (t - t_0)\dot{c}(t_0) + c(t_0)$ for any fixed $t_0 \in I$. Conversely, if $c(t) = (t - t_0)v + v_0$ then, by assumption, $1 = |\dot{c}(t)| = |v|$, and so $|\kappa(t)| = |\ddot{c}(t)| = 0$. □

1.4.3 Proposition (Characterization of the circle). *For plane curves, the following conditions are equivalent.*
  i) $|\kappa(t)| = 1/r = $ constant $ > 0$.
  ii) *c is a piece of circular arc, i.e., there exists an* $x_0 \in \mathbb{R}^2$ *with* $|c(t) - x_0| = r = $ constant $ > 0$ *for all* $t \in I$.

PROOF. We may assume $|\dot{c}(t)| = 1$.
The Frenet equations, if we assume (i), look like

$$\dot{c}(t) = e_1(t)$$
$$\dot{e}_1(t) = \epsilon/r e_2(t) \quad \text{with } \epsilon = +1 \text{ or } \epsilon = -1$$
$$\dot{e}_2(t) = -\epsilon/r e_1(t).$$

Therefore $(c(t) + \epsilon r e_2(t))^{\cdot} = \dot{c}(t) - e_1(t) = 0$, which implies that $c(t) + \epsilon r e_2(t) = x_0$, a constant vector in $\mathbb{R}^2$. Hence $c(t) - x_0 = -\epsilon r e_2(t)$, implying $|c(t) - x_0|^2 = r^2$, which is (ii).

Conversely, assume (ii). We have $(c(t) - x_0) \cdot (c(t) - x_0) = r^2$, a constant. Differentiating yields

$$\dot{c}(t) \cdot (c(t) - x_0) = 0.$$

Since $\dot{c}(t) = e_1(t)$, we have established that $c(t) - x_0$ is a multiple of $e_2(t)$. Since we know its length is r,

$$c(t) - x_0 = \epsilon r e_2(t), \quad \text{where } \epsilon = \pm 1.$$

Differentiating this equation yields

$$e_1(t) = \dot{c}(t) = \epsilon r \dot{e}_2(t) = -\epsilon r \kappa(t) e_1(t).$$

Thus $|\kappa(t)| = 1/r$. □

1.5 Space Curves

In this section we will look at curves $c: I \to \mathbb{R}^3$. In order to use Frenet-frames we assume that $\dot{c}(t)$ and $\ddot{c}(t)$ are linearly independent. By (1.2.2) we know that, under these conditions, a distinguished Frenet-frame exists.

Remark. Note that we have excluded straight lines from our consideration!

1.5.1 Definition. For a curve $c: I \to \mathbb{R}^3$, the curvatures $\kappa_1(t)$ and $\kappa_2(t)$ defined

1 Curves

in (1.3.3) will be denoted $\kappa(t)$ and $\tau(t)$ and called the "curvature" and "torsion" of c, respectively. Explicitly,

$$\kappa(t) := \frac{\dot{e}_1(t)\cdot e_2(t)}{|\dot{c}(t)|} > 0$$

$$\tau(t) := \frac{\dot{e}_2(t)\cdot e_3(t)}{|\dot{c}(t)|}.$$

The Frenet equations, in matrix form, are

$$\dot{e}(t) = |\dot{c}(t)| \begin{pmatrix} 0 & \kappa(t) & 0 \\ -\kappa(t) & 0 & \tau(t) \\ 0 & -\tau(t) & 0 \end{pmatrix} e(t).$$

1.5.2 Proposition. *If $c(t)$ is parameterized by arc length, then*

$$\kappa(t) = |\ddot{c}(t)| \quad \text{and} \quad \tau(t) = \det(\dot{c}(t), \ddot{c}(t), \dddot{c}(t))/\kappa^2(t).$$

PROOF. We know that $\dot{c}(t) = e_1(t)$, $e_2(t) = \ddot{c}(t)/|\ddot{c}(t)|$, and $e_3(t) = e_1(t) \times e_2(t) = \dot{c}(t) \times \ddot{c}(t)/|\ddot{c}(t)|$ ("$\times$" denotes the cross-product in $\mathbb{R}^3$). Thus $\kappa(t) = |\ddot{c}(t)|$, which implies $\ddot{c}(t) = \kappa(t)e_2(t)$. The Frenet equations imply

$$\begin{aligned}
\dddot{c}(t) &= \dot{\kappa}(t)e_2(t) + \kappa(t)\dot{e}_2(t) \\
&= \dot{\kappa}(t)e_2(t) + \kappa(t)[-\kappa(t)e_1(t) + \tau(t)e_3(t)] \\
&= \dot{\kappa}(t)e_2(t) - \kappa^2(t)e_1(t) + \kappa(t)\tau(t)e_3(t).
\end{aligned}$$

The equation for $\tau(t)$ now follows directly from the equations for $\dot{c}(t)$, $\ddot{c}(t)$, and $\dddot{c}(t)$ above. $\square$

Remark. By (1.3.2), $\kappa(t)$ and $\tau(t)$ are invariant with respect to isometries of $\mathbb{R}^3$ and orientation-preserving changes of variables.

Since $c(t)$ is a differentiable curve, we may write it in terms of its Taylor series at $t = t_0$. Doing so, and using the Frenet equations as they appear in (1.5.1) and (1.5.2), we get

1.5.3 Proposition (Normal (local) representation for a space curve). *Suppose $c: I \to \mathbb{R}^3$ is a space curve parameterized by arc length, and let $t_0 \in I$. Then*

$$\begin{aligned}
c(t) - c(t_0) &= \left((t - t_0) - \frac{(t - t_0)^3}{6}\kappa^2(t_0)\right)e_1(t_0) \\
&+ \left(\frac{(t - t_0)^2}{2}\kappa(t_0) + \frac{(t - t_0)^3}{6}\dot{\kappa}(t_0)\right)e_2(t_0) \\
&+ \left(\frac{(t - t_0)^3}{6}\kappa(t_0)\tau(t_0)\right)e_3(t_0) + o(t - t_0)^3.
\end{aligned}$$

1.5 Space Curves

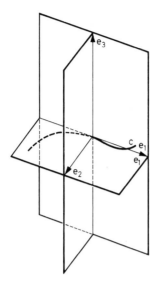

Figure 1.5

The proof follows from substituting the Frenet equations into the Taylor series.

At $t_0 \in I$ the planes in Figure 1.5 have descriptive names:

(e_1, e_2)-plane = osculating plane at $c(t_0)$.

(e_2, e_3)-plane = normal plane at $c(t_0)$.

(e_3, e_1)-plane = rectifying plane at $c(t_0)$.

Using Proposition (1.5.3), we may write down expansions for the projection of $c(t)$ onto these planes.

1.5.4 Corollary. *Let $c: I \to \mathbb{R}^3$ be a space curve parametrized by arc length and let $t_0 = 0 \in I$. Set $e_i(0) = e_i$, $1 \le i \le 3$, and $\kappa(0) = \kappa$, $\tau(0) = \tau$, $\dot{\kappa}(0) = \dot{\kappa}$. Then the projections of $c(t)$ onto the*

$$\begin{cases} \text{osculating plane at } c(t_0) \\ \text{normal plane at } c(t_0) \\ \text{rectifying plane at } c(t_0) \end{cases}$$

have Taylor expansions at 0 of the form

$$\begin{cases} \left(t, \dfrac{t^2}{2}\kappa\right) + o(t^2) \\ \left(\dfrac{t^2}{2}\kappa + \dfrac{t^3}{6}\dot{\kappa}, \dfrac{t^3}{6}\kappa\tau\right) + o(t^3) \\ \left(t - \dfrac{t^3}{6}\kappa^2, \dfrac{t^3}{6}\kappa\tau\right) + o(t^3) \end{cases}$$

1 Curves

Remark. The osculating plane derives its name from the Latin osculari, "to kiss." It is the plane spanned by the first and second derivatives of $c(t)$ at t_0 and may be thought of as the plane that fits best to $c(t)$ at $c(t_0)$. Notice that when $c(t)$ is projected onto this plane the result is, up to second order, the graph of a parabola.

The normal plane is literally that; the unique plane normal to $e_1(t_0)$, and hence to $\dot{c}(t_0)$, at $c(t_0)$.

The rectifying plane is the plane perpendicular to the "curvature vector" κe_2. Projection onto this plane "straightens" or rectifies $c(t)$ in the sense that, up to second order, the projected curve is a line.

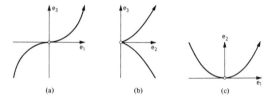

Figure 1.6 Projection onto: (a) rectifying plane; (b) normal plane; (c) osculating plane

1.6 Exercises

1.6.1 Determine the curvature of the ellipse $(a \cos t, b \sin t)$, $t \in \mathbb{R}$, $ab \neq 0$.

1.6.2 Show that the curvature of a plane curve is in general given by the formula

$$\kappa(t) = \frac{\det(\dot{c}(t), \ddot{c}(t))}{|\dot{c}(t)|^3}.$$

1.6.3 Show that the curvature and torsion of a space curve are in general given by the formulae

$$\kappa(t) = \frac{|\dot{c}(t) \times \ddot{c}(t)|}{|\dot{c}(t)|^3}$$

$$\tau(t) = \frac{\det(\dot{c}(t), \ddot{c}(t), \dddot{c}(t))}{|\dot{c}(t) \times \ddot{c}(t)|^2},$$

where $x \times y$ is the cross-product in $\mathbb{R}^3$.

1.6.4 i) Determine the curvature and torsion of the "elliptical helix"

$$(a \cos t, b \sin t, ct), \qquad ab \neq 0, \ t \in \mathbb{R}.$$

ii) Use (i) to conclude that if $a = b = 1$ then κ goes to zero as c goes to infinity. Does this make geometric sense?

Plane Curves: Global Theory

2

2.1 The Rotation Number

2.1.1 Definition. A curve $c: I = [a, b] \to \mathbb{R}^n$ is *closed* if there exists a curve $\tilde{c}: \mathbb{R} \to \mathbb{R}^n$ with the following properties: $\tilde{c} \mid I = c$ and, for all $t \in \mathbb{R}$, $\tilde{c}(t + \omega) = \tilde{c}(t)$, where $\omega = b - a$.

The number ω is *the period of c*. The curve $\tilde{c}$ is said to be *periodic with period* ω. Given a closed curve c, it is clear that its associated periodic curve $\tilde{c}$ is unique.

Remark. An equivalent definition of a closed curve is: a curve $c: [a, b] \to \mathbb{R}^n$ such that $c(a) = c(b)$ and $c^{(k)}(a) = c^{(k)}(b)$ for all $k > 0$.

For later applications we use the following generalization.

2.1.2 Definition. A *piecewise smooth curve* is a continuous function $c: [a, b] \to \mathbb{R}^n$ together with a partition

$$a = b_{-1} = a_0 < b_0 = a_1 < \ldots < b_{k-1} = a_k < b_k = a_{k+1} = b$$

of $[a, b]$ such that $c_j := c \mid [a_j, b_j]$, $0 \leq j \leq k$, is a differentiable curve. The points $c(a_j) = c(b_{j-1})$ are called *corners* of c. We will use the following terminology for piecewise smooth curves c: c is

> *regular* if each c_j is regular,
> *closed* if $c(a) = c(b)$,
> *simple closed* if c is closed and $c|_{[a,b)}$ is one-to-one.

Given a regular curve $c: I \to \mathbb{R}^2$, there is an induced map $e_1: I \to \mathbb{R}^2$, where $e_1(t) = \dot{c}(t)/|\dot{c}(t)|$, the unit tangent vector. This is sometimes called

21

the *tangent mapping*, and its image lies in $S^1 = \{x \in \mathbb{R}^2 \mid |x| = 1\}$. We begin our study of the tangent mapping by introducing a global version of the function θ considered in (1.4.1).

2.1.3 Proposition. *Let $c: [a, b] \to \mathbb{R}^2$ be a regular curve. Then there exists a continuous, piecewise differentiable function $\theta: [a, b] \to \mathbb{R}$ such that*

$$e_1(t) = \dot{c}(t)/|\dot{c}(t)| = (\cos \theta(t), \sin \theta(t)).$$

Moreover, the difference $\theta(b) - \theta(a)$ is independent of the choice of θ.

PROOF. Choose a partition $a = t_0 < t_1 < \ldots < t_k = b$ fine enough to insure that $e_1|_{[t_{j-1}, t_j]}$ lies entirely in some open semicircle of S^1. This is clearly possible since e_1 is continuous. Choose $\theta(a)$ satisfying the requirements of the proposition. Then θ is uniquely determined on $[a, t_1] = [t_0, t_1]$ by the requirement that it be continuous. If θ is known on $[t_0, t_{j-1}]$, it has a unique continuous extension to $[t_0, t_j]$; namely, $\theta(t_{j-1})$ is given and there is a unique continuous function $\tilde{\theta}: [t_{j-1}, t_j] \to \mathbb{R}$, with $\tilde{\theta}(t_{j-1}) = \theta(t_{j-1})$, satisfying the requirements of the proposition. Using $\tilde{\theta}$, we may extend θ so that it is continuous on $[t_0, t_j]$. By this procedure, θ may be defined to be continuous on $[a, b]$.

The differentiability of $\theta \mid [t_{j-1}, t_j]$ follows from (1.4.1), or directly from the differentiability of e_1 and the inverse trigonometric functions.

Finally, suppose θ and ϕ are two functions satisfying the requirements of the proposition. Then $\phi(t) - \theta(t) = 2\pi k(t)$, where $k(t)$ is a continuous integer valued function. This forces $k(t)$ to be a constant. Therefore

$$\theta(b) - \theta(a) = \phi(b) - \phi(a). \qquad \square$$

The next proposition is a technical result which will allow us to associate an "angular" function θ to a continuous mapping $e: T \to \mathbb{R}^2$, $T \subset \mathbb{R}^2$, when T is star-shaped.

2.1.4 Proposition. *Let $T \subset \mathbb{R}^2$ be star-shaped with respect to $x_0 \in T$; i.e., if $x \in T$ then the line segment $\overline{xx_0}$ is also in T. Suppose $e: T \to S^1$ is a continuous function. Then there is a continuous function $\theta: T \to \mathbb{R}$ satisfying*

$$e(x) = (\cos \theta(x), \sin \theta(x)).$$

Moreover, if θ and $\tilde{\theta}$ are two such functions, they must differ by a constant multiple of 2π.

PROOF. Choose $\theta(x_0)$ to satisfy $e(x_0) = (\cos \theta(x_0), \sin \theta(x_0))$. We may use the procedure of the proof of (2.1.3) to determine θ uniquely on each ray $\overline{x_0 x}$, $x \in T$, as a continuous function with initial value $\theta(x_0)$. What remains to be shown is that θ is continuous at any $y_0 \in T$. We may choose $\delta > 0$ such that for any $y' \in \overline{x_0 y_0}$, $|y - y'| < \delta$ implies that the angular separation between $e(y)$ and $e(y')$ is strictly less than π. Since $\overline{x_0 y_0}$ is compact and e is continuous, such a δ must exist.

Given $\epsilon > 0$, choose a neighborhood U of y_0 small enough to guarantee $U \subset B_\delta(y_0)$ and $y \in U \Rightarrow |\theta(y) - \theta(y_0)| = 2\pi k + \epsilon'$, where $|\epsilon'| < \epsilon$ and k is some integer which depends on y. Continuity of e assures the existence of such a set U. We will show $k = 0$, which implies the continuity of θ at y_0.

Let $y \in U$. Consider $\phi(s) = \theta(x_0 + s(y - x_0)) - \theta(x_0 + s(y_0 - x_0))$, $0 \le s \le 1$. ϕ is the difference between the values of θ at corresponding points on the line segments $\overline{x_0 y}$ and $\overline{x_0 y_0}$. ϕ is continuous since θ is a continuous function on each line segment.

Since $|(x_0 + s(y - x_0)) - (x_0 + s(y_0 - x_0))| = |s(y - y_0)| < \delta$, the angular separation between $(x_0 + s(y - x_0))$ and $(x_0 + s(y_0 - x_0))$ can never be equal to π. Therefore $|\phi(s) - \phi(0)| < \pi$. But $\phi(0) = 0$. Let $s = 1$, then

$$\pi > |\phi(1)| = |\theta(y) - \theta(y_0)| = |2\pi k + \epsilon'|.$$

This implies $k = 0$. $\square$

2.1.5 Definition. Let $c: [0, \omega] \to \mathbb{R}^2$ be a piecewise smooth, regular, closed curve. Let $0 = b_{-1} = a_0 < b_0 = a_1 < \ldots < b_k = \omega$ partition $[0, \omega]$ into intervals $I_j := [a_j, b_j]$ on which $c_j := c|_{I_j}$ are differentiable, $1 \le j \le k$. Let α_j denote the oriented angle from $\dot{c}(b_{j-1}) := \dot{c}(b_{j-1}-)$ to $\dot{c}(a_j) := \dot{c}(a_j+)$. The α_j, $1 \le j \le k$ are the *exterior angles* of c. We will require $-\pi < \alpha_j < \pi$.
The number

$$n_c := \frac{1}{2\pi} \sum_j (\theta_j(b_j) - \theta_j(a_j)) + \frac{1}{2\pi} \sum_j \alpha_j$$

is the *rotation number of c*.

Here the functions $\theta_j: I_j \to \mathbb{R}$, $0 \le j \le k$, are those defined in (2.1.3).

Remarks. If c is a smooth closed curve, then all $\alpha_j = 0$ and

$$n_c = \frac{\theta(\omega) - \theta(0)}{2\pi}.$$

The connection between n_c and the winding number of c as defined in elementary complex analysis is that n_c is the winding number, with respect to the origin, of the closed curve $e_1(t)$, $t \in [0, \omega]$.

EXAMPLES. i) If c is the parameterization in the positive sense (counter-clockwise) of a nondegenerate triangle, the three differentiable arcs, c_j, of which c is composed, are line segments. Therefore $\theta_j = $ constant and $\sum_{j=1}^{3} \alpha_j = 2\pi$. Hence $n_c = 1$.

Similarly, if c is a parameterization of a convex polygon, $n_c = \pm 1$.

ii) Let c be a parameterization in the positive sense of the unit circle, which makes m revolutions:

$$c(t) = (\cos 2\pi t, \sin 2\pi t), \quad 0 \le t \le m.$$

Then $n_c = m$.

2 Plane Curves: Global Theory

Notice that, in the examples above, n_c is an integer. The next proposition establishes that n_c is always an integer, and that $|n_c|$ is invariant under isometries of $\mathbb{R}^n$ and change of variables.

2.1.6 Proposition. *The rotation number n_c of a closed piecewise smooth curve is an integer. Moreover,*

(*) $$n_c = \frac{1}{2\pi} \sum_j \int_{I_j} \kappa(t)|\dot{c}(t)|\, dt + \frac{1}{2\pi} \sum_j \alpha_j.$$

As a consequence of (*) (together with (1.3.2) and the change of variables formula), n_c is invariant under orientation-preserving change of variables or congruences of $\mathbb{R}^n$. An orientation-reversing change of variables or a symmetry of $\mathbb{R}^n$ will change the sign of n_c.

PROOF. The formula defining n_c may be rewritten as

$$2\pi n_c = \sum_{j=0}^{k} (\theta_{j-1}(b_{j-1}) - \theta_j(a_j) + \alpha_j),$$

where θ_{-1} is interpreted as θ_k. By definition of α_j, $(\theta_{j-1}(b_{j-1}) - \theta_j(a_j) + \alpha_j)/2\pi$ is an integer. By (1.4.1), $\dot{\theta}_j(t) = \kappa(t)|\dot{c}(t)|$. This implies (*). □

2.2 The Umlaufsatz

The theorem we shall prove in this section is best known by its German name "Umlaufsatz." (Umlauf means "rotation" in German; Umlaufzahl = "rotation number," Satz = "theorem.")

2.2.1 Theorem (Umlaufsatz). *Let $c: I \to \mathbb{R}^2$ be a piecewise smooth, regular, simple closed plane curve. Suppose the exterior angles α_j of c are never equal to π in absolute value. Then $n_c = \pm 1$.*

2.2.2 Corollary. *Let $c: I \to \mathbb{R}^2$ be a smooth, regular, simple closed plane curve with $|\dot{c}(t)| = 1$. Then*

$$\frac{1}{2\pi} \int_I \kappa(t)\, dt = \pm 1.$$

PROOF (*due to H. Hopf*)[1]

Step 1. We will first perform a change of variables of c and an isometry of $\mathbb{R}^n$ in order to put c in a particular form. (Recall that, by (2.1.6), $|n_c|$ is invariant.)

Let g be a straight line in $\mathbb{R}^2$ which intersects the image of c. At least one point p in the intersection of g with the image of c will have the following

---

[1] Hopf, H. Über die Drehung der Tangenten und Sehnen ebener Kurven. *Compositio Math.*, **2**, 50–62 (1935).

2.2 The Umlaufsatz

property: a half-line of g with endpoint p will have no other points in common with the image of c. By performing a slight translation of g, if necessary, we can insure that p is not a corner of c (the corners of c are isolated). Thus, without loss of generality, we may assume that there is a half-line, H, emanating from a regular value, p, of c, and that H has no other points in common with the image of c. Let h be the unit vector in the direction of H.

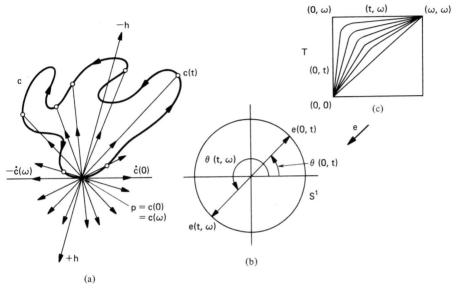

Figure 2.1 (Adapted from Manfredo P. do Carmo, *Differential Geometry of Curves and Surfaces*, Prentice-Hall, Inc., 1976, p. 396.)

Since c is regular, we may (re)parameterize c by arc length: $|\dot{c}(t)| = 1$. We also require $c(0) = c(\omega) = p$. If necessary, translation and rotation of $\mathbb{R}^2$ yields $c(0) = $ the origin and $\dot{c}(0) = e_1(0) = e_1 = (1, 0)$.

Step 2. Let $0 < a_1 < \ldots < a_{k-1} < \omega$ be a partition of $[0, \omega]$ such that c is smooth on each segment. The corners of c are the points $c(a_j)$, $0 < j < k$. Define

$$T = \{(t_1, t_2) \in \mathbb{R}^2 \mid 0 \leq t_1 \leq t_2 \leq \omega\} \backslash \{(t_1, t_2) \in \mathbb{R}^2 \mid t_1 = t_2 = a_j\}.$$

The set T is star-shaped with respect to $(0, \omega)$ (for definition, see (2.1.4)). Let $e: T \to S^1$ be the mapping defined by

$$e(t_1, t_2) = \begin{cases} \dot{c}(t_1), & \text{if } t_1 = t_2 \neq a_j, \\ -\dot{c}(0), & \text{if } (t_1, t_2) = (0, \omega), \\ \dfrac{c(t_2) - c(t_1)}{|c(t_2) - c(t_1)|}, & \text{otherwise } (t_1, t_2) \in T. \end{cases}$$

e is a continuous function (easy exercise). By Proposition (2.1.4), there exists a continuous function $\theta: T \to \mathbb{R}$ satisfying

$$(\cos \theta(t_1, t_2), \sin \theta(t_1, t_2)) = e(t_1, t_2), \quad (t_1, t_2) \in T.$$

25

2 Plane Curves: Global Theory

θ is determined up to an integral multiple of 2π. We choose θ to satisfy $\theta(0, \omega) = +\pi$.

Step 3. We will show that $\theta(\omega, \omega) - \theta(0, 0) = \pm 2\pi$. For $t \in]0, \omega[$, $\theta(t, \omega) - \theta(0, \omega)$ measures the angle between $-e_1$ and the unit vector

$$e(t, \omega) = \frac{c(\omega) - c(t)}{|c(\omega) - c(t)|}.$$

But $e(t, \omega)$ can never be equal to $-h$. Therefore $\theta(t, \omega) - \theta(0, \omega)$ is always less than 2π. So when $t = \omega$, $\theta(\omega, \omega) - \theta(0, \omega) = \pm \pi$.

Similarly, $\theta(0, t) - \theta(0, 0)$, which represents the angle from e_1 to $e(0, t)$, is equal to 0 when $t = 0$ and can never exceed 2π. Therefore as $t \to \omega$, $\theta(0, t) - \theta(0, 0) \to \pm \pi$. The sign here is the same as that of $\theta(\omega, \omega) - \theta(0, \omega)$. Thus $\theta(\omega, \omega) - \theta(0, 0) = \theta(\omega, \omega) - \theta(0, \omega) + \theta(0, \omega) - \theta(0, 0) = \pm 2\pi$.

Step 4. Consider $c(a_j) = c(b_{j-1})$, a corner of c with exterior angle α_j. The angle α_j is equal to the angle between $\dot{c}(b_{j-1})$ and $\dot{c}(a_j)$, measured in the positive sense. Define

$$\theta(a_j, a_j) = \lim_{t \to a_j} \theta(t, t), \qquad t > a_j,$$

$$\theta(b_{j-1}, b_{j-1}) = \lim_{t \to b_{j-1}} \theta(t, t), \qquad t < b_{j-1}.$$

Claim: $\alpha_j = \theta(a_j, a_j) - \theta(b_{j-1}, b_{j-1})$.

PROOF. Let Δ be the triangle whose vertices are $x_{-1} := c(b_{j-1} - \epsilon)$, $x_0 := c(b_{j-1}) = c(a_j)$, $x_1 := c(a_j + \epsilon)$, where ϵ satisfies $b_{j-2} < b_{j-1} - \epsilon < b_{j-1} + \epsilon < b_j$. Assume that x_{-1}, x_0, x_1 orders the vertices of Δ in the positive sense. Without loss of generality, Δ may be assumed to be nondegenerate. Let $\alpha_{\pm 1}$, $0 < \alpha_{\pm 1} < \pi$, be the angle at vertex $x_{\pm 1}$. Then $\theta(b_{j-1}, a_j + \epsilon) - \theta(b_{j-1} - \epsilon, a_j + \epsilon) = \alpha_1 + 2\pi k_1$ for some integer k_1. If ϵ is chosen small enough, $\theta(b_{j-1}, a_j + \epsilon) - \theta(t, a_j + \epsilon)$, $b_j - \epsilon \le t \le b_{j-1}$, cannot exceed 2π, so $k_1 = 0$. Similarly, $\theta(b_{j-1} - \epsilon, a_j + \epsilon) - \theta(b_{j-1} - \epsilon, a_j) = \alpha_{-1}$. Therefore $\theta(a_j, a_j + \epsilon) - \theta(b_{j-1} - \epsilon, b_{j-1}) = \alpha_1 + \alpha_{-1} = \pi - \beta_0$, where β_0 is the angle at x_0. As $\epsilon \to 0$, $\pi - \beta_0 \to \alpha_j$, the exterior angle of $c(t)$ at a_j. This proves the claim.

If x_{-1}, x_0, x_1 orients Δ in the negative direction, an analogous proof will work.

Step 5. Conclusion of proof of theorem. By Steps 3 and 4, we may write

$$\pm 2\pi = \theta(\omega, \omega) - \theta(0, 0)$$

$$= \theta(\omega, \omega) - \sum_{j=1}^{k-1} \theta(a_j, a_j) + \sum_{1}^{k-1} \theta(b_{j-1}, b_{j-1}) + \sum_{j=1}^{k-1} \alpha_j - \theta(0, 0).$$

Since $\theta(a_j, a_j) = \theta_j(a_j)$ and $\theta(b_{j-1}, b_{j-1}) = \theta_j(b_{j-1})$ as defined in (2.1.5), the right-hand side is $2\pi n_c$. Here we have $\omega = b_k$, $0 = a_0$. This proves the theorem.

Step 6. Proof of corollary. The corollary follows immediately from the theorem and (2.1.6). □

2.3 Convex Curves

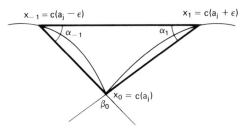

Figure 2.2

2.3 Convex Curves

2.3.1 Definition. A regular plane curve $c: I \to \mathbb{R}^2$ is *convex* if, for all $t_0 \in I$, the curve lies entirely on one side of the tangent at $c(t_0)$. In other words, for every $t_0 \in I$ one and only one of the following inequalities hold:

$$(c(t) - c(t_0)) \cdot e_2(t_0) \geq 0, \quad \text{all } t \in I$$

or

$$(c(t) - c(t_0)) \cdot e_2(t_0) \leq 0, \quad \text{all } t \in I.$$

2.3.2 Theorem (A characterization of convex curves). *Let $c: I \to \mathbb{R}^2$ be a simple closed regular plane curve. Then c is convex if and only if one of the following conditions are true*:

$$\kappa(t) \geq 0, \quad \text{all } t \in I$$

or

$$\kappa(t) \leq 0, \quad \text{all } t \in I.$$

Remarks. i) If one of the above conditions hold then an orientation-reversing change of variables will produce the other. So, geometrically, they are equivalent.
ii) If c is closed but not simple, the theorem fails. For example, a trefoil (pretzel curve) satisfies $\kappa(t) \geq 0$, but it is not convex.

Figure 2.3

PROOF. *Step 1.* We may assume, without loss of generality, that (after possibly a change of variables) $|\dot{c}(t)| = 1$. If we then consider the function $\theta: I \to \mathbb{R}$, defined in (2.1.3), we may assert that $\dot{\theta}(t) = \kappa(t)$. This is proved in (1.4.1).

Step 2. Suppose c is convex. We will show that κ does not change sign by showing that $\theta(t)$ is weakly monotone. If $\theta(t') = \theta(t'')$ and $t' < t''$ then θ is constant on $[t', t'']$.

First observe that since c is simple, there must be at least one point t''' where $\theta(t''') = -\theta(t'') = -\theta(t')$. Using the convexity of c, it is possible to conclude that two of the tangent lines to c at the points $c(t')$, $c(t'')$, $c(t''')$ must coincide.

Let $p_1 = c(t_1)$ and $p_2 = c(t_2)$, $t_1 < t_2$, denote these two points, and consider the line segment $\overline{p_1 p_2}$. This line segment must lie entirely on the image of c. For suppose q is a point on $\overline{p_1 p_2}$ which is not on the image of c. The line perpendicular to $\overline{p_1 p_2}$ and through q intersects c in at least two points and, since c is convex, these points must lie on the same side of $\overline{p_1 p_2}$. Let r (resp. s) be the points of intersection closest to (resp. furthest from) $\overline{p_1 p_2}$. Then r lies in the interior of the triangle $p_1 p_2 s$. Consider the tangent line to c at the point r. Whatever it is, there are points of c on both sides of it, contradicting the fact that c is convex. Hence $\overline{p_1 p_2} = \{c(t) \mid t_1 \leq t \leq t_2\}$, which means that $\theta(t_1) = \theta(t) = \theta(t_2)$ for $t \in [t_1, t_2]$. In particular, $t_1 = t'$ and $t_2 = t_2''$. This concludes the proof of weak monotonicity.

Step 3. Suppose c is *not* convex. This means there exists a $t_0 \in I$ such that $\phi(t) := (c(t) - c(t_0)) \cdot e_2(t_0)$ changes sign. Let t_+ and t_- ($\neq t_0$) be values of $t \in I$ where $\phi(t)$ assumes its maximum and minimum, respectively:

$$\phi(t_-) < \phi(t_0) = 0 < \phi(t_+).$$

Since $\dot\phi(t_-) = \dot\phi(t_+) = 0$, $e_1(t_+)$ and $e_1(t_-) = \pm e_1(t_0)$. Therefore at least two of these vectors are equal. By reparameterizing, we may now assert that there exist s_1, s_2, with $s_1 = 0 < s_2 < \omega$ and

$$e_1(s_1) = e_1(s_2).$$

But this means that $\theta(s_2) - \theta(s_1) = 2\pi k$, k an integer, and $\theta(s_1 + \omega) - \theta(s_2) = 2\pi k'$, k' an integer. By the Umlaufsatz, $k + k' = \pm 1$ and, since $\theta|_{[0,s_2]}$ and $\theta|_{[s_2,\omega]}$ are nonconstant functions, $kk' \neq 0$. Therefore $kk' < 0$, which means that $\kappa(t) = \dot\theta(t)$ must change sign (one of the "k"'s is positive, the other negative). This completes the proof. □

We will now use this characterization of a convex curve to prove the well-known *four vertex theorem.*

2.3.3 Definition. A *vertex* of a smooth plane curve $c: I \to \mathbb{R}^2$ is a critical point of the curvature $\kappa: I \to \mathbb{R}$ in the interior $\mathring{I}$ of I, i.e., $\dot\kappa(t_0) = 0$, $t_0 \in \mathring{I}$. If $\kappa(t) = \text{const}$, $t_1 \leq t \leq t_2$, all these t are vertices.

2.3.4 Theorem (Four vertex theorem). *A convex, simple, closed smooth plane curve has at least four vertices.*

Remark. The theorem is true without the convexity hypothesis (although it is harder to prove).

2.4 Exercises and Some Further Results

PROOF (*due to G. Herglotz*)[2]

Step 1. Since $\kappa(t)$ has a maximum and a minimum on I, $c(t)$ has at least two vertices. Without loss of generality, we may assume that c is parameterized by arc length and that $\kappa(t)$ has a minimum at $t = 0$ and a maximum at t_0, $0 < t_0 < \omega$, where $I = [0, \omega]$. After a suitable rotation, we may also assume that the line through $c(0)$ and $c(t_0)$ is the x-axis in the (x, y) plane, and that, if $c(t) = (x(t), y(t))$, there exists at least one point $\hat{t}$, $0 < \hat{t} < t_0$, with $y(\hat{t}) > 0$. (If $y(t) \equiv 0$, $0 \leq t \leq t_0$, then $\kappa(t) \equiv 0$, $0 \leq t \leq t_0$, implying $\kappa \equiv 0$ on I, an impossibility.)

Step 2. Claim: $c(0)$ and $c(t_0)$ are the only points of c on the x-axis. For if $c(t_1)$ is another point of c on the x-axis, the convexity of c forces the tangent line to $c(t)$ at the middle point of $c(0)$, $c(t_0)$, $c(t_1)$ to pass through the other two points. As in the proof of (2.32), this implies that the line segment $\overline{c(0)c(t_0)}$ lies entirely in the image of c, making $\kappa(0) = \kappa(t_0) = 0$. This is impossible since it would imply $\kappa(t) \equiv 0$ on I.

Step 3. Suppose $c(t_0)$ and $c(0)$ are the only vertices of c. Then

$$\dot{\kappa}(t) \geq 0 \quad \text{for } t \in [0, t_0]$$
$$\dot{\kappa}(t) \leq 0 \quad \text{for } t \in [t_0, \omega].$$

This implies that $\dot{\kappa}(t) y(t) \geq 0$ for $t \in [0, \omega]$. Therefore

$$0 \leq \int_0^\omega \dot{\kappa}(t) y(t) \, dt = -\int_0^\omega \kappa(t) \dot{y}(t) \, dt,$$

using integration by parts.

Since $e_1(t) = (\dot{x}(t), \dot{y}(t))$, $\dot{e}_1(t) = \kappa(t) e_2(t)$ and $e_2(t) = (-\dot{y}(t), \dot{x}(t))$, it follows that $\ddot{x}(t) = -\kappa(t) \dot{y}(t)$. Therefore

$$0 \leq \int_0^\omega \dot{\kappa}(t) y(t) \, dt = -\int_0^\omega \kappa(t) \dot{y}(t) \, dt = \int_0^\omega \ddot{x}(t) \, dt = 0.$$

This can only be true if $y(t) \equiv 0$, so we have arrived at a contradiction.

Step 4 (conclusion). We have actually shown that, under the hypotheses, there must be another point t where $\dot{\kappa}(t)$ changes sign, i.e., where κ has a relative extremum. Relative extrema come in pairs; so there must be at least four vertices. □

2.4 Exercises and Some Further Results

2.4.1 A convex curve $c: I \to \mathbb{R}^2$ with $\kappa(t) \neq 0$ for all $t \in I = [0, \omega]$ is said to be *strictly convex*.

Prove: If c is a closed, strictly convex curve, then for every $v \in S^1$ there exists a unique $t \in I$ such that $e_1(t) = v$.

---

[2] See Blaschke [A2], pp. 31–32, or Chern [A6], pp. 23–25.

2.4.2 By (2.4.1), for every point $c(t)$ on a closed, strictly convex curve $c: I \to R^2$, there is a unique point $c(t')$ such that $e_1(t) = -e_1(t')$. c is said to have *constant width* if $d(c(t), c(t')) = d$, a constant.

Prove: The circumference of a closed, strictly convex curve of constant width $= d$ is equal to πd.

2.4.3 If a closed, strictly convex curve c has exactly four vertices, then any circle has at most four points of intersection with c.[3]

2.4.4 If a closed, strictly convex curve intersects a circle in $2n$ points, then it has at least $2n$ vertices.[3]

2.4.5 The four vertex theorem can be derived from the following result concerning closed curves c in $\mathbb{R}^3$ with no self-intersections. Suppose c is strictly convex, in the sense that through each point of c there passes a plane which has no other points in common with c. Then c has at least four points with stationary osculating plane; i.e., four points $c(t)$ where $\tau(t) = 0$. For a proof of this result, see Barner.[4]

2.4.6 A closer look at our proof of the four vertex theorem will show that we may actually claim a stronger result: a simple closed convex curve must have either $\kappa \equiv$ constant $\neq 0$ or a curvature function κ with two relative maxima and two relative minima. In the latter case, we may also require that the values of κ at the relative maxima be strictly greater than the values of κ at the relative minima.

From this theorem we see that not every periodic $\kappa: [0, \omega] \to \mathbb{R} \geq 0$ occurs as the curvature function of a closed convex curve $c: I \to \mathbb{R}^2$. It turns out that the necessary restrictions on κ given above are also sufficient.

Theorem (a converse to the four vertex theorem) (Gluck).[5] Let $\kappa: [0, \omega] \to \mathbb{R} > 0$ be a continuous, strictly positive, periodic function ($\kappa(0) = \kappa(\omega)$) which is either constant or has two maxima and two minima, the values of κ at the maxima being strictly greater than the values of κ at the minima. Then there exists a C^2 curve $c: [0, \omega] \to \mathbb{R}^2$ which is simple and closed and whose curvature function is equal to κ.

The four vertex theorem (2.3.4) has the following generalization: Let c be a simple, closed, null-homotopic curve on M, an oriented surface with a Riemannian metric of constant Gauss curvature. Then the geodesic curvature of c has at least four stationary points.

If M has variable, nonpositive Gauss curvature, a version of the four-vertex theorem is still true with the same hypotheses as above, provided one generalizes the notion of a vertex to mean a point of c where c may be well approximated by a "circle of hyperbolic geometry." The meaning of this approximation can be precisely defined. In case M has constant Gauss curvature, the derivative of the geodesic curvature vanishes at these generalized vertices (Thorbergsson).[6]

---

[3] See Blaschke, Kreis, and Kugel [A4], p. 161.

[4] Barner, M. Über die Mindestanzahl stationärer Schmiegebenen bei geschlossenen strengkonvexen Raumkurven. *Abh. Math. Sem. Univ.-Hamburg*, **20**, 196–215 (1956).

[5] Gluck, H. The converse to the four vertex theorem. *L'Enseignement Mathématique*, IIe Serie, Tome XVII, 3–4 (1971), pp. 295–309.

[6] Thorbergsson, G. Vierscheitelsatz auf Flächen. *Math. Z.*, **149**, 47–56 (1976).

2.4 Exercises and Some Further Results

2.4.7 By the Jordan curve theorem, a simple closed plane curve, c, divides the plane into two disjoint regions, one of which is bounded. If L = length of c and A = area of the bounded region, then $L^2 - 4\pi A \geq 0$. Equality holds if and only if c is a circle. This is the famous *isoperimetric inequality* (proved in Chern [A5], p. 23).[7]

A stronger form of this inequality exists for closed convex curves.[8] If r is the radius of the largest disc lying inside the bounded region, or the radius of the smallest disc containing the bounded region, then $L^2 - 4\pi A \geq (A - \pi r^2)^2/r^2$. For further developments, see Osserman.[9]

2.4.8 Consider the following problem. Given $p, q \in \mathbb{R}^2$ and $X \in T_p\mathbb{R}^2$, $Y \in T_q\mathbb{R}^2$, unit vectors, find the curve of shortest length from p to q with initial direction X and final direction Y. A solution does not always exist; let $p \neq q$ and $X \perp Y$. However, if the class of curves is restricted to those with "average curvature" equal to or less than $1/r$, $r > 0$, and C^1 (but possibly not C^2) curves are allowed, then a solution always exists. In fact, the solution curves consist of circular arcs and line segments. Moreover, there are, at most, three different arcs of this type on any solution curve. This result is due to L. E. Dubins.[10]

2.4.9 Corollary (2.2.2) of the Umlaufsatz can be generalized to closed curves $c: I \to \mathbb{R}^n$, $n \geq 3$. Recall that, for $n > 2$, $\kappa > 0$ for the curves we considered in Section (1.5). The *total curvature* of c is defined as

$$K(c) = \int_0^L |\kappa(t)|\, dt,$$

where c is assumed to be parameterized by arc length.

Theorem (Fenchel[11]). $K(c) \geq 2\pi$, with equality, if and only if c is a convex plane curve.

This theorem was generalized by Fary and Milnor.[12] They proved that if $c: I \to \mathbb{R}^3$ is closed and knotted, then $K(c) \geq 4\pi$. A curve c is knotted if no homeomorphism of $\mathbb{R}^3$ will move c onto the unit circle in the (x, y) plane. Equivalently, c is knotted if it does not bound an embedded disc in $\mathbb{R}^3$.

---

[7] An early proof of the isoperimetric inequality, although not one which completely satisfies today's mathematical standards, was given by J. Steiner: Steiner, J. Einfache Beweise der isoperimetrischen Hauptsätze. *J. Reine Angew. Math.* **18**, 289–296 (1838).

[8] Bonneson, T. *Les problèmes des isopérimètres et des isépiphanes*. Gauthier-Villars, Paris, 1929.

[9] Osserman, R. Isoperimetric and related inequalities. *Proc. AMS Symp. in Pure and Applied Math.* **XXVII**, Part 1, 207–215.

[10] Dubins, L. E. On curves of minimal length with constraint on average curvature and prescribed initial and terminal positions and tangents. *Amer. J. Math.*, **79**, 497–516 (1957).

[11] Fenchel, W. Über Krümmung und Wendung geschlossener Raumkurven. *Math. Ann.* **101**, 238–252 (1929). Cf. also Fenchel, W. On the differential geometry of closed space curves. *Bull. Amer. Math. Soc.*, **57**, 44–54 (1951), or Chern [A5].

[12] Fary, I. Sur las courbure totale d'une courbe gauche faisant un noeud. *Bull. Soc. Math. France*, **77**, 128–138 (1949). Milnor, J. On the differential geometry of closed space curves. *Ann. of Math.*, **52**, 248–257 (1950).

2.4.10 *A proof of Fenchel's theorem.* i) Prove: Let $c: I \to \mathbb{R}^n$ be a closed curve lying on $S^{n-1}(r) = \{x \in \mathbb{R}^n \mid |x| = r\}$, i.e., $|c(t)| = r$, $t \in [0, \omega]$. Suppose c does not lie in any open hemisphere of $S^{n-1}(r)$. Then the length of c is at least $2\pi r$. (A simple proof of this is given by Horn[13].) Using this result,

ii) Prove: Fenchel's theorem (2.4.9).

[13] Horn, R. A. On Fenchel's theorem. *Amer. Math. Monthly*, **78**, 380–381 (1971).

Surfaces: local theory 3

3.1 Definitions

3.1.1 Definitions. i) U will always denote an open set in $\mathbb{R}^2$. Points of U will be denoted by $u \in \mathbb{R}^2$, or by $(u^1, u^2) \in \mathbb{R} \times \mathbb{R}$ or $(u, v) \in \mathbb{R} \times \mathbb{R}$.
ii) A differentiable mapping $f: U \to \mathbb{R}^3$ such that $df_u: T_u\mathbb{R}^2 \to T_{f(u)}\mathbb{R}^3$ is injective for all $u \in U$ is a (*parameterized*) *surface patch*, or simply a *surface*. A mapping f satisfying this condition is called *regular*. The $u \in U$ are called *parameters* of f.
iii) The two-dimensional linear subspace $df_u(\mathbb{R}_u^2) \subset T_{f(u)}\mathbb{R}^3$ is called *the tangent space of f at u*, and will be denoted by $T_u f$. Elements of $T_u f$ are called *tangent vectors* (of f at u).

3.1.2 Examples. i) $f(u, v) = x_0 + ux + vy$, where x, y are linearly independent vectors in $\mathbb{R}^3$. The map $f: U \to \mathbb{R}^3$ parameterizes a piece of a plane.
ii) $U = D^2 = \{(u, v) \in R^2 \mid u^2 + v^2 < 1\}, f(u, v) = (u, v, \sqrt{1 - u^2 - v^2})$.

The map f parameterizes a hemisphere.

Remark. The natural basis $e_1 = (1, 0), e_2 = (0, 1)$ of $T_u\mathbb{R}^2 \simeq \mathbb{R}^2$ is mapped by df_u into a basis of $T_u f$. We shall write $df_u e_1 = (\partial f / \partial u^1)(u^1, u^2)$, $df_u e_2 = (\partial f / \partial u^2)(u^1, u^2)$ or simply $df_u(e_1) = f_{u^1}, df_u(e_2) = f_{u^2}$, where $u = (u^1, u^2)$.

These basis vectors of $T_{u_0} f \subset T_{f(u_0)} \mathbb{R}^3 = \mathbb{R}^3$ are equal to the first partial derivatives of f at (u_0^1, u_0^2): since

$$|f(u) - f(u_0) - df_{u_0}(u - u_0)| = o(u - u_0),$$

3 Surfaces: Local Theory

this implies

$$\lim_{u^1 \to u_0^1} \left| \frac{f(u) - f(u_0)}{u^1 - u_0^1} - df_{u_0}(1, 0) \right| = 0 \quad \text{with } u_0 = (u_0^1, u_0^2),\ u = (u^1, u_0^2).$$

Therefore $df_{u_0}(1, 0) = df_{u_0} e_1$ is equal, in coordinates, to $(\partial f/\partial u^1)(u_0^1, u_0^2)$. Similarly, $df_{u_0}(0, 1) = (\partial f/\partial u^2)(u_0^1, u_0^2)$.

3.1.3 Definition. Let $f\colon U \to \mathbb{R}^3$ be a surface. A *change of variables* of f is a diffeomorphism $\phi\colon V \subset \mathbb{R}^2 \to U \subset \mathbb{R}^2$, where V is an open set in $\mathbb{R}^2$, such that $d\phi$ always has rank $= 2$. If $\det(d\phi) > 0$, ϕ is *orientation preserving*. The surface $\tilde{f} := f \circ \phi\colon V \to \mathbb{R}^3$ is said to be *related to f by the change of variables ϕ*.

Remark. Relationship by change of variables defines an equivalence relation on the class of all surfaces. An equivalence class of mappings is called an *unparameterized surface*.

3.1.4 Definition. A *vector field along* $f = \{f\colon U \to \mathbb{R}^3\}$ is a differentiable mapping $X\colon U \to \mathbb{R}^3$.

We think of a vector field X along f as taking values in the tangent space of $\mathbb{R}^3$ restricted to the surface f, i.e., $X(u) \in T_{f(u)}\mathbb{R}^3$. To make this explicit, consider the map

$$\tilde{X}\colon U \to T\mathbb{R}^3 \quad \text{given by } u \mapsto (f(u), X(u)).$$

$\tilde{X}$ is clearly a differentiable mapping and, for a given f, determines the mapping X. Conversely, given a vector field X along a map f, we usually interpret it as defining the corresponding map $\tilde{X}$.

3.1.5 Definition. A vector field X along $f\colon U \to \mathbb{R}^3$ is

$$\begin{cases} \text{tangential} & \text{if } (f(u), X(u)) \in T_u f \text{ for all } u \in U, \\ \text{normal} & \text{if } (f(u), X(u)) \in T_{f(u)}\mathbb{R}^3 \text{ is orthogonal to } T_u f \text{ for all } u \in U. \end{cases}$$

For example, $f_{u^1}(u)$ and $f_{u^2}(u)$ are tangential vector fields along f. They are sometimes called the *coordinate vector fields*. The vector field of $f_{u^1}(u) \times f_{u^2}(u)$ (cross-product in $\mathbb{R}^3$) is a normal vector field along f. All three are obviously differentiable.

3.1.6 Proposition. *Every tangential vector field X along a surface $f\colon U \to \mathbb{R}^3$ may be represented in the following form*:

(*) $$X(u) = a^1(u) f_{u^1}(u) + a^2(u) f_{u^2}(u).$$

The real-valued functions $a^1(u)$ and $a^2(u)$ are differentiable and uniquely determined. Conversely, a pair of differentiable functions $a^i\colon U \to \mathbb{R}$, $i = 1, 2$, determines a unique tangential vector field of the form (*).

PROOF. The last statement is clear. Moreover, given $X(u)$, the functions $a^1(u)$ and $a^2(u)$ are uniquely determined. What remains to be shown is that the

$a^i(u)$ are differentiable. To prove differentiability, take the inner product of (*) with $f_{u^1}(u)$ and $f_{u^2}(u)$:

$$\sum_i a^i(u) f_{u^i} \cdot f_{u^k} = X(u) \cdot f_{u^k}, \quad k = 1, 2.$$

This gives a system of linear equations for $a^1(u)$, $a^2(u)$. The coefficients are differentiable functions, and $\det(f_{u^i} \cdot f_{u^j}) \neq 0$. By using Cramer's rule, one can see that the $a^i(u)$, $i = 1, 2$, are differentiable. □

3.1.7 Definition. Let $n := (f_{u^1} \times f_{u^2})/|f_{u^1} \times f_{u^2}|$. The vector field n is called the *(Gauss) unit normal field along f*. The mapping $n: U \to S^2 \subset \mathbb{R}^3$ is also referred to as the *Gauss map*. The moving 3-frame (f_{u^1}, f_{u^2}, n) is called the *Gauss frame* of the surface $f: U \to \mathbb{R}^3$.

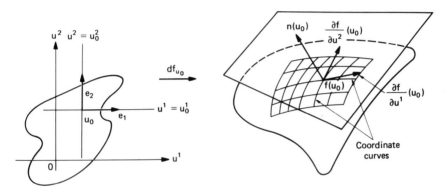

Figure 3.1 The Gauss Frame at a point of f (Adapted from Manfredo P. do Carmo, *Differential Geometry of Curves and Surfaces*, Prentice-Hall, Inc., 1976, p. 39.)

N.B. This is in general *not* an orthonormal frame.

3.2 The First Fundamental Form

3.2.1 A quick review of quadratic forms

1. Let T be a real vector-space.
    A *symmetric bilinear form* or a *quadratic form* is a map $\beta: T \times T \to \mathbb{R}$ satisfying

$$\beta(X, Y) = \beta(Y, X) \quad \text{(symmetry)}$$
$$\beta(aX + bY, Z) = a\beta(X, Z) + b\beta(Y, Z) \quad \text{(bilinearity)}.$$

Here $a, b \in \mathbb{R}$ and $X, Y, Z \in T$. β is *positive definite* if

$$X \neq 0 \Rightarrow \beta(X, X) > 0.$$

Example: The standard inner product in Euclidean space $\mathbb{R}^n$.

3 Surfaces: Local Theory

2. The *matrix representation* of β with respect to a basis e_i, $1 \leq i \leq n$, of T is the matrix

$$(g_{ij}) := (\beta(e_i, e_j)).$$

If $X = \sum_i \xi^i e_i$, $Y = \sum_j \eta^j e_j$, then $\beta(X, Y) = \sum_{i,j} \xi^i \eta^j g_{ij}$.

Suppose f_k, $1 \leq k \leq n$, is another basis of T. Let (a_i^k) be the matrix defined by $e_i = \sum_k a_i^k f_k$, $1 \leq i \leq n$. If $\beta(f_k, f_l) := h_{kl}$, then $g_{ij} = \sum_{k,l} a_i^k a_j^l h_{kl}$. If $G = (g_{ij})$, $A = (a_i^j)$, and $H = (h_i^j)$ these equations may be written in matrix form:

$$G = A \cdot H \cdot {}^t A,$$

the dot denoting matrix multiplication.

3. Let $L: S \to T$ be a linear mapping between vector spaces S and T. Suppose β is a quadratic form on T. Then, for $X, Y \in S$,

$$\alpha(X, Y) := \beta(LX, LY)$$

defines a quadratic form on S. The form α is said to be *induced by β via L*. If L is injective and β is positive definite, then α is positive definite. Suppose $X \neq 0$. Then $LX \neq 0$ and $\alpha(X, X) = \beta(LX, LX) > 0$.

3.2.2 Definitions. i) Let $f: U \to \mathbb{R}^3$ be a surface. Let $u \in U$. The inner product on $\mathbb{R}^3 \cong T_{f(u)}\mathbb{R}^3$ induces a quadratic form on $T_u f \subset T_{f(u)}\mathbb{R}^3 \cong \mathbb{R}^3$ by restriction. This form is called the *first fundamental form* and is denoted sometimes by g or g_u and sometimes by I or I_u.

ii) The inner product on $T_{f(u)}\mathbb{R}^3 \cong \mathbb{R}^3$ composed with the linear map $df_u: \mathbb{R}^2 \cong T_u \mathbb{R}^2 \to T_{f(u)}\mathbb{R}^3 \cong \mathbb{R}^3$ induces a quadratic form on $T_u \mathbb{R}^2$ which is also called the first fundamental form. It is also denoted by g or I, and it will sometimes be written "$df \cdot df$."

Remark. The linear bijection $df_u: T_u\mathbb{R}^2 \cong \mathbb{R}^2 \to \mathbb{R}^2 \cong T_u f$ is clearly an isometry with respect to the first fundamental form, i.e.,

$$I_u(df_u X, df_u Y) = I_u(X, Y) \quad \text{for } X, Y \in T_u \mathbb{R}^2.$$

Therefore, if we identify $T_u \mathbb{R}^2$ with $T_u f$ by means of df_u, we may identify these two definitions of the first fundamental form. Once more: For X and Y in $T_u \mathbb{R}^2$, $I(X, Y) := df_u X \cdot df_u Y$. For X and Y in $T_u f$, $I(X, Y) := X \cdot Y$.

3.2.3 Definition. The matrix representation of the first fundamental form, with respect to the basis f_{u^1}, f_{u^2}, will be denoted by

$$(g_{ik}) := (g(f_{u^i}, f_{u^k})).$$

Sometimes we will use the notation $E := g(f_{u^1}, f_{u^1})$, $F := g(f_{u^1}, f_{u^2}) = g(f_{u^2}, f_{u^1})$, $G := g(f_{u^2}, f_{u^2})$ (Gauss' notation). Here $g_{ik}(u) = f_{u^i}(u) \cdot f_{u^k}(u)$. By the definition of I on $T_u \mathbb{R}^2$ in (3.2.2), $(g_{ik}(u))$ is also equal to the matrix representation $(I(e_i, e_k))$ of I with respect to the canonical basis e_1, e_2 of $T_u \mathbb{R}^2$.

3.2 The First Fundamental Form

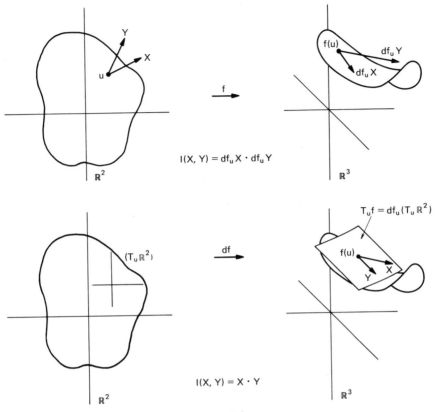

Figure 3.2

3.2.4 Proposition. i) *The first fundamental form I of a surface $f: U \to \mathbb{R}^3$ is positive definite.*

ii) *I is differentiable, i.e., the coefficients of the matrix $g_{ik}: U \to \mathbb{R}$ are differentiable. This is equivalent to the following condition: For any $X: U \to \mathbb{R}^3$, $Y: U \to \mathbb{R}^3$, tangential vector fields along f, the map $u \to g_u(X(u), Y(u))$ is differentiable.*

PROOF. i) follows from (3.2.1, 3).

ii) follows from the definition of $g_{ik}(u) = f_{u^i}(u) \cdot f_{u^k}(u)$ and from the last statement of Proposition (3.1.6). □

3.2.5 Proposition (Invariance of the first fundamental form). *Let $f: U \to \mathbb{R}^3$ be a surface.*

i) *Let $B: \mathbb{R}^3 \to \mathbb{R}^3$ be an isometry. Then $\tilde{f} := B \circ f$ is also a surface and*
$$\tilde{I}_u(dBX, dBY) = I_u(X, Y) \quad \text{for all } X, Y \in T_u f.$$

ii) *Let $\phi: V \to U$ be a change of variables and let $\tilde{f} = f \circ \phi$. Then*
$$\tilde{I}_v(X, Y) = I_{\phi(v)}(X, Y) \quad \text{for all } X, Y \in T_v \tilde{f} = T_{\phi(v)} f$$

and
$$\tilde{I}_v(\tilde{X}, \tilde{Y}) = I_{\phi(v)}(d\phi \tilde{X}, d\phi \tilde{Y}) \quad \text{for all } X, Y \in T_v\mathbb{R}^2.$$

PROOF. i) Suppose $Bx = Rx + x_0$, where R is an orthogonal map. Then $dB = R$. Therefore, if $X, Y \in T_u f$, $\tilde{I}_u(RX, RY) = RX \cdot RY = X \cdot Y = I_u(X, Y)$.

ii) Let $\tilde{X}, \tilde{Y} \in T_v\mathbb{R}^2$. Then $\tilde{I}_v(\tilde{X}, \tilde{Y}) = d\tilde{f}_v\tilde{X} \cdot d\tilde{f}_v\tilde{Y} = df_u \circ d\phi \tilde{X} \cdot df_u \circ d\phi \tilde{Y} = I_u(d\phi \tilde{X}, d\phi \tilde{Y})$, where $u = \phi(v)$. □

3.2.6 Corollary. *Suppose the change of variables ϕ is given, in terms of coordinates, by $u^i = u^i(v^1, v^2)$, $i = 1, 2$. Then the fundamental matrix $(\tilde{g}_{ij})$ of $\tilde{f} = f \circ \phi$ is related to the fundamental matrix (g_{ij}) of f by*

$$\tilde{g}_{ij}(v) = \sum_{k,l} \frac{\partial u^k}{\partial v^i} \frac{\partial u^l}{\partial v^j} g_{kl}(\phi(v)).$$

PROOF. $d\phi \tilde{e}_i = \sum_k (\partial u^k/\partial v^i) e_k$, where $(\tilde{e}_i)$ (respectively (e_k)) is the canonical basis of $T_v\mathbb{R}^2$ (respectively $T_{\phi(v)}\mathbb{R}^2$). The corollary follows by applying the formula (3.2.5, (ii)) to $\tilde{X} = \tilde{e}_i$, $\tilde{Y} = \tilde{e}_j$. *Note*: Since we know that $d\phi_v \colon T_v\mathbb{R}^2 \to T_{\phi(v)}\mathbb{R}^2$ has the matrix representation $(\partial u^k/\partial v^i)$ (see Chapter 0), we may prove the corollary by using (3.2.1, 2). Specifically, if $A = (\partial u^k/\partial v^i)$, then $(\tilde{g}_{ij}) = \tilde{G} = A \cdot G \cdot {}^t\!A$, where $G = (g_{ij})$. □

3.3 The Second Fundamental Form

3.3.1 Definition. Let $f \colon U \to \mathbb{R}^3$ be a surface. The map
$$n \colon U \to S^2 \subset \mathbb{R}^3, \quad u \mapsto n(u)$$
is called the *Gauss map*. In words, n maps u into the unit normal vector $n(u)$ to f at $f(u)$. Each $n(u)$ lies in $T_{f(u)}\mathbb{R}^3$. By using the canonical identification of $T_{f(u)}\mathbb{R}^3$ with $\mathbb{R}^3$, we may consider n as a mapping from U to $\mathbb{R}^3$.

Remark. Since $n(u)$ is a unit vector, $n(u) \in S^2 \subset \mathbb{R}^3$, where
$$S^2 = \{x \in \mathbb{R}^3 \mid |x| = 1\}.$$

3.3.2 Proposition. *The image of $dn_u \colon T_u\mathbb{R}^2 \to T_{f(u)}\mathbb{R}^3$ lies in $T_u f \subset T_{f(u)}\mathbb{R}^3$.*

PROOF. $dn_u(T_u\mathbb{R}^2) = $ span of n_{u^1}, n_{u^2}. Since $n(u) \cdot n(u) = 1$, differentiation yields $n_{u^i}(u) \cdot n(u) = 0$, $i = 1, 2$. This means $n_{u^i} \in T_u f$. Here we have canonically identified $T_{n(u)}\mathbb{R}^3$ with $T_{f(u)}\mathbb{R}^3$. □

3.3.3 Proposition. *The mapping*
$$(X, Y) \in T_u\mathbb{R}^2 \times T_u\mathbb{R}^2 \mapsto -dn_u X \cdot df_u Y \in \mathbb{R}$$
is a symmetric bilinear form on $T_u\mathbb{R}^2$.

PROOF. Bilinearity is obvious. To prove symmetry, observe that, since $n \cdot f_{u^j} = 0$,
$$-n_{u^i} \cdot f_{u^j} = n \cdot f_{u^i u^j}$$
$$= n \cdot f_{u^j u^i} = -n_{u^j} \cdot f_{u^i}. \qquad \square$$

3.3 The Second Fundamental Form

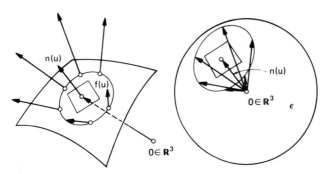

Figure 3.3 The Gauss map

3.3.4 Definition. i) The quadratic form

$$-dn_u \cdot df_u \colon T_u\mathbb{R}^2 \times T_u\mathbb{R}^2 \to \mathbb{R}$$

is called the *second fundamental form of f at u*, and is denoted by II (or II_u).

ii) The linear mapping $L_u := -dn_u \circ df_u^{-1} \colon T_u f \to T_u f$ is called the *Weingarten map*.

Remarks. 1. $II_u \colon T_u f \times T_u f \to \mathbb{R}$ can be written as $L_u X \cdot Y$, i.e.,

$$II_u(X, Y) = L_u X \cdot Y \quad \text{for all } X, Y \in T_u f.$$

2. The matrix representation of II_u with respect to the canonical basis $\{e_i\}$ of $T_u\mathbb{R}^2$ and the associated basis $\{f_{u^i}\}$ of $T_u f$ is

$$(h_{ik}) := (-n_{u^i} \cdot f_{u^k}) = (n \cdot f_{u^i u^k}).$$

Sometimes we will use Gauss' notation:

$$\begin{pmatrix} L & M \\ M & N \end{pmatrix} := \begin{pmatrix} h_{11} & h_{12} \\ h_{21} & h_{22} \end{pmatrix}.$$

See section 3.7 for examples.

3.3.5 Definition. The *third fundamental form of f at u* is the symmetric bilinear form given by

$$(X, Y) \in T_u\mathbb{R}^2 \times T_u\mathbb{R}^2 \mapsto dn_u X \cdot dn_u Y \in \mathbb{R}.$$

The third fundamental form is denoted by III_u, III or $dn \cdot dn$. If we want to consider III as a form on $T_u f$, it is given by $L_u X \cdot L_u Y$.

Proposition (3.2.5) for I has a counterpart for II:

3.3.6 Proposition. *II is invariant (in the sense of (3.2.5)) under congruences of $\mathbb{R}^3$ and orientation-preserving changes of variables.*

3 Surfaces: Local Theory

PROOF. i) Let $Bx = Rx + x_0$ be a congruence (det $R = 1$). Then $\tilde{f} = B \circ f$ is a surface and $\tilde{f}_{u^i} = dBf_{u^i} = Rf_{u^i}$, $\tilde{n} = dBn$. Therefore, if $X, Y \in T_u f$,

$$\tilde{II}_u(dBX, dBY) = -d\tilde{n} \circ d\tilde{f}_u^{-1}(dBX) \cdot dBY$$
$$= -dB(dn \circ df_u^{-1}(X)) \cdot dBY = -dn \circ df_u^{-1}X \cdot Y = II_u(X, Y).$$

ii) Let $\phi: V \to U$ be an orientation-preserving change of variables and $\tilde{f} = f \circ \phi$. Then $\tilde{f}_{v^k} = \sum_i f_{u^i} \, \partial u^i / \partial v^k$, and this implies that

$$\tilde{f}_{v^1} \times \tilde{f}_{v^2} = (f_{u^1} \times f_{u^2}) \det\left(\frac{\partial u^i}{\partial v^k}\right).$$

Therefore $\tilde{n} = n \circ \phi$, since $\det(\partial u^i / \partial v^k) > 0$. Thus, for $X, Y \in T_u \tilde{f}$, we have

$$\tilde{II}_v(X, Y) = -d\tilde{n} \circ d\tilde{f}^{-1}(X) \cdot Y = -dn \circ df^{-1}(X) \cdot Y = II_{\phi(v)}(X, Y)$$

and, for $\tilde{X}, \tilde{Y} \in T_v \mathbb{R}^2$, we have

$$\tilde{II}_v(\tilde{X}, \tilde{Y}) = -dn \circ d\phi \tilde{X} \cdot df \, d\phi \tilde{Y} = II_u(d\phi \tilde{X}, d\phi \tilde{Y}) \quad \text{with } u = \phi(v). \quad \square$$

3.3.7 Examples

1. *The sphere*

$$f(u, v) := (\cos u \cdot \cos v, \cos u \cdot \sin v, \sin u), \quad (u, v) \in {]-\pi/2, \pi/2[} \times \mathbb{R}.$$

The image of f is S^2 minus the north and south poles: $S^2 - \{0, 0, \pm 1\}$

$$f_u = (-\sin u \cos v, -\sin u \sin v, \cos u)$$
$$f_v = (-\cos u \sin v, \cos u \cos v, 0)$$

$E = f_u^2 = g_{11} = 1, \qquad F = f_u \cdot f_v = g_{12} = 0, \qquad G = f_v^2 = g_{22} = \cos^2 u.$

$$n(u, v) = \frac{(f_u \times f_v)}{|f_u \times f_v|} = -(\cos u \cos v, \cos u \sin v, \sin u)$$
$$= -f(u, v)$$
$$II = -dn \cdot df = df \cdot df = I.$$

2. *The torus*

$$g(u, v) := ((a + b \cdot \cos u) \cos v, (a + b \cdot \cos u) \sin v, b \cdot \sin u),$$
$$0 < b < a, (u, v) \in \mathbb{R} \times \mathbb{R}.$$

$$g_u = b(-\sin u \cos v, -\sin u \sin v, \cos u)$$
$$g_v = (a + b \cdot \cos u)(-\sin v, \cos v, 0)$$
$$g_{11} \equiv E = g_u^2 = b^2, \qquad g_{12} \equiv F = g_u \cdot g_v = 0$$
$$g_{22} \equiv G = g_v^2 = (a + b \cdot \cos u)^2$$
$$n(u, v) = -(\cos u \cos v, \cos u \sin v, \sin u).$$

3.3 The Second Fundamental Form

Thus, for $u \in\,]-\pi/2, \pi/2[$, $n(u, v) = -f(u, v)$ where f is as in (1) above.

$$II = -dn \cdot dg = df \cdot dg$$
$$h_{11} \equiv L = f_u \cdot g_u = b$$
$$h_{12} \equiv M = f_u \cdot g_v + f_v \cdot g_u = 0$$
$$h_{22} \equiv N = f_v \cdot g_v = (a + b \cos u) \cos u.$$

$$\det(h_{ik}) = b \cos u (a + b \cos u) \text{ is } \begin{cases} > 0 & \text{for } -\dfrac{\pi}{2} < u < \dfrac{\pi}{2} \\ = 0 & \text{for } u = \pm\dfrac{\pi}{2} \\ < 0 & \text{for } \dfrac{\pi}{2} < u < \dfrac{3\pi}{2}. \end{cases}$$

These three cases are the outside, the top and bottom circle, and the inside, respectively.

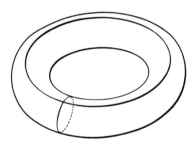

Figure 3.4 Torus

3. *Surfaces of revolution*

$$f(u, v) := (h(u) \cos v, h(u) \sin v, k(u))$$

where $h'^2 + k'^2 \neq 0$, $h \neq 0$. The surface parameterized by f is the surface generated by revolving the curve $(h(u), 0, k(u))$ about the z-axis

$$f_v^2 = h'^2 + k'^2, \qquad f_u \cdot f_v = 0, \qquad f_u^2 = h^2.$$

4. *Surfaces generated by one-parameter groups of isometries*

A *one-parameter group of isometries* of $\mathbb{R}^3$ is a differentiable mapping $\gamma: \mathbb{R} \times \mathbb{R}^3 \to \mathbb{R}^3$ with the following properties:

The map $\gamma_t: \mathbb{R}^3 \to \mathbb{R}^3$ given by $x \to \gamma(t, x)$, $(t, x) \in \mathbb{R} \times \mathbb{R}^3$, is an isometry, $\gamma_t \circ \gamma_s = \gamma_{t+s}$ and $\gamma_0 =$ the identity.

It may be shown that, possibly after a change of basis, any one-parameter group may be written as

$$\gamma(t, x) = (x^1 \cos t + x^2 \sin t, -x^1 \sin t + x^2 \cos t, x^3 + bt).$$

3 Surfaces: Local Theory

The *orbit* $t \in \mathbb{R} \mapsto \gamma(t, x) \in \mathbb{R}^3$ of a point $(x^1, x^2, x^3) = x$ which does not lie on the x^3-axis is a helix (see Example 2 in 1.1). A *generated surface* is a surface produced by a curve $c(v)$, $v \in J$, and a one-parameter group of isometries γ:

$$f(u, v) = \gamma(u, c(v)), \quad (u, v) \in I \times J.$$

It is certainly possible that f is not a regular map, so one needs to assume additional conditions to insure that f is a surface. Some examples of generated surfaces are the sphere, the torus, and, more generally, any surface of revolution.

An example of a generated surface which is not a surface of revolution is given by the *helicoid*. Let $c(v) = (v, 0, 0)$, $v \in \mathbb{R}$, and let

$$\gamma(t, x) = (x^1 \cos t + x^2 \sin t, -x^1 \sin t + x^2 \cos t, x^3 + bt), \quad b \neq 0, t \in \mathbb{R}.$$

Then the generated surface

$$f(u, v) = \gamma(u, c(v)) = (v \cos u, -v \sin u, bu)$$

is in fact a surface in the sense of (3.1). Moreover:

$$f_u = (-v \sin u, -v \cos u, b)$$
$$f_v = (\cos u, -\sin u, 0)$$

$$g_{ij} = \begin{pmatrix} b^2 + v^2 & 0 \\ 0 & 1 \end{pmatrix}$$

$$n = \frac{(b \sin u, b \cos u, v)}{(b^2 + v^2)^{1/2}}$$

$$n_u = \frac{b(\cos u, -\sin u, 0)}{(b^2 + v^2)^{1/2}}$$

$$h_{ij} = \begin{pmatrix} 0 & \frac{-b}{(b^2 + v^2)^{1/2}} \\ \frac{-b}{(b^2 + v^2)^{1/2}} & 0 \end{pmatrix}$$

$$n_v = \frac{b}{(b^2 + v^2)^{3/2}}\left(-r \sin u, -v \cos u, b\right)$$

The helicoid may be thought of as the surface generated by a ray perpendicular to the z-axis which is rotating at a constant speed in the plane parallel to the (x, y) plane and moving at a constant speed in the z-direction.

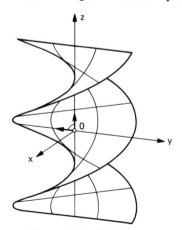

Figure 3.5 Helicoid

3.4 Curves on Surfaces

3.4.1 Definition. Let $f: U \to \mathbb{R}^3$ be a surface. By a *curve on f* we mean a curve $c: I \to \mathbb{R}^3$ which can be written in the form $f \circ u$, where $u: I \to U \subset \mathbb{R}^2$ is a curve in U.

The study of curves on surfaces will give us a geometric interpretation of the first and second fundamental forms.

3.4.2 Proposition. *Let $c = f \circ u: I \to \mathbb{R}^3$ be a curve on f. Then*
$$\dot{c}(t) = \sum_i \dot{u}^i f_{u^i} \circ u(t)$$
is a tangent vector to f at $u(t)$. The length of $\dot{c}(t)$ is given by
$$|\dot{c}(t)|^2 = \sum_{i,j} g_{ij}(u(t)) \dot{u}^i(t) \dot{u}^j(t).$$

PROOF. $\dot{c}(t) = dc_t(1) = df_{u(t)} \circ du_t(1) = df_{u(t)}(\sum_i \dot{u}^i e_i) = \sum_i \dot{u}^i f_{u^i} \circ u(t)$. The desired formula follows from the definition $|\dot{c}(t)|^2 = \langle \dot{c}(t), \dot{c}(t) \rangle$. □

Remarks. For a curve $c(t) = f \circ u(t)$ on f, the arc-length parameter $s(t)$ is uniquely determined by the following formula:
$$\left(\frac{ds}{dt}\right)^2 = |\dot{c}(t)|^2 = \sum_{i,j} g_{ij} \frac{du^i}{dt} \frac{du^j}{dt} = I(\dot{u}, \dot{u}).$$

The first fundamental form may be expressed, in terms of this notation, as
$$ds^2 = \sum_{i,j} g_{ij} \, du^i \, du^j = I(du, du).$$

The expression ds is called the *line element* of the surface f.

Suppose $c = f \circ u: I \to \mathbb{R}^3$ is a unit-speed curve on $f: U \to \mathbb{R}^3$ for which $\dot{c}(t), \ddot{c}(t)$ are linearly independent. The curve $c(t)$ possesses a distinguished Frenet-frame $(e_1(t), e_2(t), e_3(t))$ and the curvature of c is defined by the equation $\dot{e}_1(t) = \kappa(t) e_2(t)$ (see (1.5.1)). The relationship between the curvature of c and the second fundamental form of f is given by the following proposition.

3.4.3 Proposition. *Let $c = f \circ u$ be a curve which satisfies the hypotheses in the above remark. Then*
$$II(\dot{c}(t), \dot{c}(t)) = \kappa(t) n(t) \cdot e_2(t)$$
with $n(t) = n \circ u(t)$.

Corollary (Meusnier's theorem). *Let $\theta(t) \in [0, \pi/2]$ be the angle between the normal to f and the osculating plane of c (i.e., $\theta = \sphericalangle(n, e_3)$). Then*
$$|II(\dot{c}(t), \dot{c}(t))| = \kappa(t) \cos \theta(t).$$

Consequently, if $\theta(t) < \pi/2$,

$$\kappa(t) = \frac{|II(\dot{c}(t), \dot{c}(t))|}{\cos \theta(t)}.$$

PROOF. $II(\dot{c}(t), \dot{c}(t)) = -dn(\dot{u}(t)) \cdot df(\dot{u}(t)) = -\dot{n}(t) \cdot \dot{c}(t)$. Since $n(t) \cdot \dot{c}(t) = 0$, this implies

$$II(\dot{c}(t), \dot{c}(t)) = n(t) \cdot \ddot{c}(t) = \kappa(t) n(t) \cdot e_2(t). \qquad \square$$

Remark. For an arbitrary curve, the above results are *not* true.

3.4.4 Definitions. i) Let $X \in T_u f$, $|X| = 1$ be a unit tangent vector on a surface f. The *normal curvature in the direction* $\pm X$ is the number

$$\kappa(X) = \kappa(-X) := II(X, X).$$

ii) Let $c = f \circ u \colon I \to \mathbb{R}^3$ be a unit-speed curve on f for which $\dot{c}, \ddot{c}$ are also linearly independent. Let $(e_i(t))$ be the distinguished Frenet-frame of c at t. If $e_2(t_0) = \pm n(c(t_0))$, c is said to lie in a *normal section* at $t = t_0$.

Remark. If $II(X, X) \neq 0$ for some $X \in T_u f$, then $|\kappa(X)|$ is equal to the curvature $\kappa(t_0)$ of a curve c at $c(t_0) = f \circ u(t_0)$ which lies in a normal section at $t = t_0$. Of course, we assume that $\dot{c}(t_0)$ and $\ddot{c}(t_0)$ are linearly independent. By hypothesis, $\kappa(t_0) > 0$. Therefore $II(X, X)$ is positive or negative, depending on whether $e_2(t_0)$ is equal to plus or minus $n(u(t_0))$ = the unit normal vector to the surface at $u(t_0)$.

3.4.5 Examples. We will continue those examples introduced in (3.3.7).

1. *The sphere.* Clearly $II = I$ and $\kappa(X) = 1$ for all X. The requirement that $c(t)$ be a normal section at each point forces $c(t)$ to be a great circle.

2. *The torus.* Consider a typical meridian circle on the torus, e.g., $c(t) = g(t, 0) = (a + b \cos t, 0, b \sin t)$. This is a circle with radius b, curvature $\kappa(t) = 1/b$, $e_1(t) = (-\sin t, 0, \cos t)$, $e_2(t) = -(\cos t, 0, \sin t)$.

Using the expression for $n(u(t))$ computed in (3.3.7), we see that $n(u(t)) = e_2(t)$. Therefore $\kappa(e_1(t)) = 1/b$.

For the inner and outer equators of the torus,

$$c(t) = ((a \pm b) \cos t, (a \pm b) \sin t, 0),$$

a simple computation will show

$$e_1(t) = (-\sin t, \cos t, 0)$$

$$e_2(t) = (-\cos t, -\sin t, 0)$$

$$\kappa(t) = 1/(a \pm b)$$

$$n(t) = \{\mp\}(\cos t, \sin t, 0) = \begin{cases} e_2(t) & \text{for } u = 0, \\ -e_2(t) & \text{for } u = \pi. \end{cases}$$

3.5 Principal Curvature, Gauss Curvature, Mean Curvature

Therefore

$$\kappa(e_1(t)) = \begin{cases} \dfrac{1}{a+b} & \text{on inner equator,} \\ -\dfrac{1}{a-b} & \text{on outer equator.} \end{cases}$$

3.4.6 Definition. Let $c: I \to \mathbb{R}^3$ be a space curve with the property that $\dot{c}(t)$ and $\ddot{c}(t)$ are linearly independent. The *osculating circle of c at t* is the circle with radius $1/\kappa(t)$ lying in the plane of $e_1(t), e_2(t)$ with center $c(t) + e_2(t)/\kappa(t)$.

Remark. The osculating circle is characterized by the following property. It is the limit as $t', t'' \to t$ of the circle passing through the points $c(t'), c(t)$, and $c(t'')$. (Proof: exercise.)

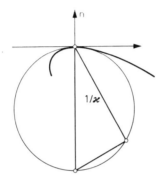

Figure 3.6 Osculating circle

The relationship between the local behavior of $c(t)$ and the osculating circle is given by the next proposition, a typical result of classical surface theory.

3.4.7 Proposition. *Suppose f is a surface and X is a tangent vector at u_0 with $|X| = 1$ and $\kappa(X) \neq 0$. If $c = f \circ u$ is a curve on f with $\dot{c}(t_0) = \pm |\dot{c}(t_0)| X$, then the osculating circle of c at t_0 is the intersection of the osculating plane of c at t_0 with the sphere of radius $1/|\kappa(X)|$ centered at $f(u_0) + n(u_0)/\kappa(X)$.*

PROOF. By Meusnier's theorem, (3.4.3), $1/\kappa(t_0) = n(t_0) \cdot e_2(t_0)/\kappa(X)$, provided $e(t_0) \cdot n(t_0) \neq 0$. In other words, $c(t_0) + e_2(t_0)/\kappa(t_0)$, which is the center of the osculating circle, is equal to the projection of the vector $n(t_0)/\kappa(X)$ in the $e_2(t_0)$ direction. □

3.5 Principal Curvature, Gauss Curvature, and Mean Curvature

3.5.1 Definition. Let $f: U \to \mathbb{R}^3$ be a surface. Let

$$S_u^1 f := \{ X \in T_u f \mid I_u(X, X) = 1 \}$$

denote the unit circle in $T_u f$. A vector $X_0 \in S_u^1 f$ is said to be a *principal*

direction if X_0 is a critical point of the function
$$X \in S_u^1 f \mapsto \kappa(X) = II(X, X) \in \mathbb{R}.$$
If X_0 is a principal direction, the value $\kappa(X_0)$ is called a *principal curvature of f at u.*

Note that if X is a principal direction, so is $-X$. There are always at least two linearly independent principal directions, namely the values of X where $\kappa(X)$ takes on a maximum and a minimum on the compact set $S_u^1 f$.

The principal curvatures are characterized by the following proposition.

3.5.2 Proposition (Rodriguez). *Let $X \in S_u^1 f$. Then X is a principal direction if and only if X is an eigenvector of the Weingarten map*
$$L_u = -dn_u \circ df_u^{-1} : T_u f \to T_u f.$$
The associated eigenvalues are the principal curvatures.

PROOF. Suppose κ is a principal curvature with associated principal direction X_0, $I(X_0, X_0) = 1$. Using the Lagrange multiplier rule,[1] we may assert that $d(II - \kappa I) = 0$ at X_0, $I(X_0, X_0) = 1$. Since II and I are both quadratic forms and since the differential of any quadratic form β at a point X is given by $d\beta Y = 2\beta(X, Y)$, the above requirement is equivalent to
$$II(X_0, Y) - \kappa I(X_0, Y) = 0 \qquad I(X_0, X_0) = 1, \quad \text{for all } Y,$$
which in turn is equivalent to
$$L_u X_0 = \kappa X_0, \qquad I(X_0, X_0) = 1, \qquad \kappa = II(X_0, X_0) = \kappa(X_0).$$
Therefore X_0 is an eigenvector of L_u with eigenvalue κ. Conversely, let X_0 be an eigenvector of L_u with eigenvalue κ. Then if $X + \epsilon Y$ satisfies $I(X_0 + \epsilon Y, X_0 + \epsilon Y) = 1$,
$$II(X_0 + \epsilon Y, X_0 + \epsilon Y) - I(X_0, X_0) = 2\epsilon II(X_0, Y) + \epsilon^2(\ldots)$$
and $2\epsilon I(X_0, Y) + \epsilon^2 I(Y, Y) = 0$.

Therefore $2II(X_0, Y) = 2\kappa I(X_0, Y) = -\kappa \epsilon I(Y, Y)$ and
$$II(X_0 + \epsilon Y, X_0 + \epsilon Y) - II(X_0, X_0) = 0 + \epsilon^2(\ldots).$$
The last equation clearly implies that X_0 is a critical point of $\kappa(X)$ on $S_u^1 f$, i.e., X_0 is a principal direction. □

3.5.3 Corollary. *Either II is proportional to I ($II = \kappa I$), in which case every direction is a principal direction, or there exist exactly two (up to sign) principal directions orthogonal to each other.*

---

[1] See Edwards, C. H. *Advanced Calculus of Several Variables.* Academic Press, New York, 1973, pp. 90–99.

3.5 Principal Curvature, Gauss Curvature, Mean Curvature

PROOF. Let κ_1, κ_2 be the largest and smallest principal curvatures, respectively. If X_1 and X_2 are associated principal directions, then $\kappa_1 I(X_1, X_2) = II(X_1, X_2) = \kappa_2 I(X_1, X_2)$. Therefore either $\kappa_1 = \kappa_2$, i.e., $\kappa = $ const, and II is proportional to I or $\kappa_1 > \kappa_2$ and $I(X_1, X_2) = 0$. Suppose κ_0 is any principal curvature with principal direction X_0. Then either $I(X_0, X_1)$ or $I(X_0, X_2) = 0$, which implies that either $X_0 = \pm X_1$ or $X_0 = \pm X_2$. □

We will now use the principal curvatures to define two important functions.

3.5.4 Definition. Let $f: U \to \mathbb{R}^3$ be a surface. The *Gauss curvature* and the *mean curvature* of f are the following two functions on U:

$$K(u) := \kappa_1(u) \cdot \kappa_2(u) \qquad H(u) := \tfrac{1}{2}(\kappa_1(u) + \kappa_2(u)).$$

3.5.5 Proposition. i) *The curvature functions K and H are determined by the equation $\det(\kappa \, id + dn \circ df^{-1}) = \kappa^2 - 2H\kappa + K$, where the left-hand side is the characteristic polynomial $\chi(L_u)$ of the Weingarten map $L_u = -dn \circ df_u^{-1}$ in the variable κ. Consequently, $2H(u) = \text{Tr}L(u)$ and $K(u) = \det L(u)$.*

ii) *If (h_{ik}) is the matrix representation of II, (g_{ik}) is the matrix representation of I, and (g^{ik}) is the inverse of (g_{ik}), then L_u has the matrix representation*

(*) $$(a_i^k) = \left(\sum_j h_{ij} g^{jk}\right).$$

Consequently,

$$K(u) = \frac{\det II_u}{\det I_u} = \frac{\det(h_{ik}(u))}{\det(g_{ik}(u))}$$

$$2H(u) = \sum_{i,k} h_{ik}(u) g^{ik}(u).$$

Note: From the representations for K and H, it follows that they are differentiable functions.

PROOF. i) The principal curvatures κ_1, κ_2 are solutions to $\kappa^2 - 2H\kappa + K = 0$, the characteristic equation of $-dn \circ df^{-1}$. Therefore $\kappa^2 - 2H\kappa + K = (\kappa - \kappa_1)(\kappa - \kappa_2)$.

ii) With respect to the standard basis of $T_u f$,

$$-dn \circ df^{-1}(f_{u^i}) = -dn(e_i) = -n_{u^i} = \sum_k a_i^k f_{u^k}.$$

Taking the inner product with f_{u^j}:

$$h_{ij} = \sum_k a_i^k g_{kj},$$

which implies (*). Now $\det(\kappa \delta_i^k - \sum_j h_{ij} g^{jk}) = \kappa^2 - 2H\kappa + K$ (by (i)), from which the expressions for H and K follow directly.

It is reasonable to ask why the third fundamental form has not entered directly into our study of curvature on surfaces. It turns out that the third fundamental form is totally determined by the first and second fundamental forms. □

3.5.6 Proposition. $III - 2HII + KI = 0$.

PROOF. Let $IV := (dn + \kappa_1 df) \cdot (dn + \kappa_2 df)$, where κ_1, κ_2 are the principal curvatures. Clearly,

$$IV = III - 2HII + KI.$$

But $IV(X_1, Y) = IV(Y, X_2) = 0$, where X_i is a principal direction for κ_i, $i = 1, 2$, and Y is arbitrary. Therefore $IV = 0$. □

Remark. This proposition is a special case of Cayley's theorem: A linear mapping L (in our case, the Weingarten map) satisfies $\chi(L) = 0$, where $\chi(\kappa) = \det(L - \kappa(id))$.

The various curvature functions we have been considering are invariant under change of variables and isometries as the following theorem shows.

3.5.7 Theorem. *Let $f: U \to \mathbb{R}^3$ be a surface and $X \in T_u f$ a principal direction with associated principal curvature $\kappa = \kappa(X)$. Let $K(u)$ and $H(u)$ be the Gauss and mean curvatures, respectively.*

  i) *If $B: \mathbb{R}^3 \to \mathbb{R}^3$ is an isometry, then $\tilde{f} := B \circ f$ is also a surface and $\tilde{X} := dBX \in T_u \tilde{f}$ is a principal direction of $\tilde{f}$, $\tilde{\kappa}(\tilde{X}) = \pm \kappa(X)$, $\tilde{K}(u) = K(u)$, and $\tilde{H}(u) = \pm H(u)$. The signs are positive if B is a congruence, negative if B is a symmetry.*

  ii) *If $\phi: V \to U$ is a change of variables, then $\tilde{f} := f \circ \phi$ is a surface and $\tilde{X} := X$ is a principal direction of f and $\tilde{\kappa}(v) = \pm \kappa \circ \phi(v)$, $\tilde{K}(v) = K \circ \phi(v)$, and $\tilde{H}(v) = \pm H \circ \phi(v)$. The sign is positive if ϕ is orientation-preserving, negative if ϕ is orientation-reversing.*

PROOF. i) From the proof of (3.3.6) we see that $\tilde{n}(u) = \pm dBn(u)$, the sign depending on whether B is orientation-preserving or reversing. Therefore,

$$-d\tilde{n} \circ d\tilde{f}^{-1} \tilde{X} = \mp dB \circ dn \circ df^{-1} X = \pm \kappa \, dBX = \pm \kappa \tilde{X}.$$

This means that $\tilde{X}$ is a principal direction with principal curvature $\tilde{\kappa} = \pm \kappa$.

ii) From the proof of (3.3.6) we see that $\tilde{n}(v) = \pm n \circ \phi(v)$, the sign being positive if $\det d\phi > 0$, negative otherwise. Therefore $-d\tilde{n} \circ d\tilde{f}^{-1} \tilde{X} = \mp dn \circ df^{-1} X = \pm \kappa \tilde{X}$. This means that $\tilde{X} = X$ is a principal direction with associated principle curvature $\tilde{\kappa} = \pm \kappa$.

Remark. The Gauss curvature K is the only one of the curvature functions which does not change sign under orientation-reversing isometries or change of variables.

3.5.8 Examples. We continue the examples developed in (3.3.7) and (3.4.5).
1. On the sphere, $\kappa_1 = \kappa_2 = 1$, $H = K = 1$.
2. The torus. We compute $a_i^k = -\sum_j h_{ij} g^{jk}$, $a_2^1 = b^{-1}$, $a_1^2 = a_2^1 = 0$, $a_2^2 = \cos u/(a + b \cos u)$. Therefore $\kappa_1 = a_2^2 < \kappa_2 = a_1^1 = b^{-1}$. Compare this with (3.4.3). Also,

$$K = \cos u/b(a + b \cos u), \qquad H = (a + 2b \cos u)/2b(a + b \cos u).$$

The maximum $\kappa_2 = b^{-1}$ is assumed by any principal direction, X, which is tangential to a meridian circle: $II(X, X) = \kappa(X) = \kappa_2$.

The minimum κ_1 is not a constant function. It is positive on the outside of the torus, i.e., when $u \in \,]{-\pi/2}, \pi/2[$. It is negative on the inside of the torus, i.e., when $u \in \,]\pi/2, 3\pi/2[$. Finally, $\kappa_1 = 0$ on the top and bottom latitude circles, i.e., $u = \pm\pi/2$.

Consequently, $K > 0$ on the outside of the torus, $K < 0$ on the inside of the torus, and $K = 0$ on the top and bottom latitude circles.

3.5.9 Definition. Let $f: U \to \mathbb{R}^3$ be a surface. A point $u_0 \in U$ is called an *umbilic* if $\kappa_1(u_0) = \kappa_2(u_0)$. If, in addition, $\kappa_1(u_0) = \kappa_2(u_0) = 0$, then u_0 is said to be a *planar point*.

3.5.10 Definition. A surface $f: U \to \mathbb{R}^3$ is said to be *planar* (resp. *spherical*) if $n(u) = $ constant (respectively, if there exists an $x_0 \in \mathbb{R}^3$ such that $|f(u) - x_0| = \rho$, a positive constant).

3.5.11 Proposition. *A surface consists entirely of umbilics if and only if it is planar or spherical.*

PROOF. 1. If f is planar or spherical, then $dn = 0$ or $df \cdot (f - x_0) = 0$. The latter condition implies that $n = \pm(f - x_0)/|f - x_0| = \pm(f - x_0)/\rho$. Therefore $dn = -\kappa\, df$, where $\kappa = 0$ or $\kappa = $ constant $= \pm 1/\rho$.

2. Let $dn = -\kappa\, df$. Therefore $n_u = -\kappa f_u$, $n_v = -\kappa f_v$. Consequently, $n_{uv} = -\kappa_v f_u - \kappa f_{uv} = -\kappa_u f_v - \kappa f_{vu}$. Since f_u and f_v are linearly independent, $\kappa_u = \kappa_v = 0$, so $\kappa = $ constant. If $\kappa = 0$, $dn = 0$. Therefore $n = $ constant and f is planar. If $\kappa \neq 0$, then $((n/\kappa) + f)_u = ((n/\kappa) + f)_v = 0$. Therefore $(n/\kappa) + f = x_0$, a constant vector. Consequently, $|f - x_0| = 1/|\kappa| = \rho = $ constant, so f is spherical. □

3.6 Normal Form for a Surface, Special Coordinates

In our investigation of curves we were able to analyze local behavior by expressing the curve up to second order in terms of a Frenet-frame at a fixed point (see (1.5.3)). Here is the analog for surfaces.

3.6.1 Proposition. *Let $f: U \to \mathbb{R}^3$ be a surface, $u_0 \in U$, $\{X_1, X_2\}$ a basis of $T_{u_0} f$, and $n_0 = n(u_0)$ the unit normal at u_0 which makes $\{X_1, X_2, n_0\}$ positively* oriented. Then there is a change of variables $\phi: V_0 \to U_0 \subset U$ near u_0 with $\phi(0) = u_0$ with the following properties: if $\tilde{f} = f \circ \phi$,

$$\tilde{f}(v) - \tilde{f}(0) = v^1 X_1 + v^2 X_2 + r(v) n_0, \qquad v = (v^1, v^2).$$

If $X_i = f_{u^i}(u_0)$, then $v^i = u^i - u_0^i + o(|u - u_0|)$ and $r_{v^i v^j}(0) = h_{ij}(u_0)$.

PROOF. Since $\{X_1, X_2, n\}$ forms a basis in $T_{f(u)} \mathbb{R}^3$, we may write

$$f(u) - f(u_0) = v^1(u) X_1 + v^2(u) X_2 + q(u) n_0$$

for some functions $v^i(u), q(u)$ with $v^i(u_0) = q(u_0) = 0$. The first order of business is to find an inverse for $v = (v^1(u), v^2(u))$. Since

$$f_{u^i}(u_0) = \sum_k \frac{\partial v^k}{\partial u^i}(u_0) X_k,$$

$((\partial v^k / \partial u^i)(u_0))$ is an invertible matrix. The inverse function theorem insures the existence of a local inverse ϕ to v, defined in a neighborhood V_0 of 0. This is the change of variables we seek. For, if $\tilde{f} = f \circ \phi$,

$$\tilde{f}(v) - \tilde{f}(0) = \sum_i v^i X_i + r(v) n_0, \quad \text{where } r = q \circ \phi.$$

It is clearly seen that $\tilde{h}_{ij}(0) = r_{v^i v^j}(0)$. If the X_i happen to be $f_{u^i}(u_0)$, then $(\partial v^i / \partial u^j)(u_0) = \delta^i_j$ and $\tilde{h}_{ij}(0) = (\partial v^i / \partial u^j)(u_0)$. Therefore $(\tilde{h}_{ij}(0)) = (h_{ij}(u_0))$. □

3.6.2 Definition. A surface $f: U \to \mathbb{R}^3$ is

$$\begin{cases} \text{elliptic} \\ \text{parabolic} \\ \text{hyperbolic} \end{cases} \text{ at } u_0 \in U \text{ if det } II_{u_0} \text{ is } \begin{cases} > 0 \\ = 0 \\ < 0. \end{cases}$$

Let us assume now that our surface $f: U \to \mathbb{R}^3$ is presented in the standard form of (3.6.1) with $X_i = f_{u^i}$. Since $r(0) = r_{u^i}(0) = 0$,

$$r(u) = \frac{1}{2} \sum_{i,j} h_{ij}(0) u^i u^j + o(|u|^2).$$

Consequently,

$$f(u) - f(0) = \sum_i u^i f_{u^i}(0) + \frac{1}{2} \sum_{i,j} h_{ij}(0) u^i u^j n(0) + o(|u|^2).$$

We have "proved" the following result.

3.6.3 Proposition. If f is

$$\begin{cases} \text{elliptic} \\ \text{parabolic} \quad (\text{with } II_{u_0} \neq 0) \\ \text{hyperbolic} \end{cases}$$

3.6 Normal Form for a Surface, Special Coordinates

at u_0, then the surface represented by the second Taylor polynomial of f is an

$$\begin{cases} \text{elliptic paraboloid,} \\ \text{parabolic cylinder,} \\ \text{hyperbolic paraboloid.} \end{cases}$$

This representation gives us a geometric picture of what the sign of the Gauss curvature means, since its sign is the same as the sign of det II.

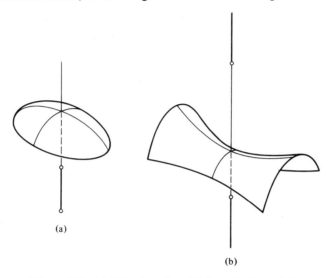

Figure 3.7 (a) Elliptic point; (b) hyperbolic point

We now turn out attention to finding coordinates on a surface fitted to vector fields that are given in advance. The basic tool is the following theorem.

3.6.4 Theorem. *Suppose X_1 and X_2 are tangential vector fields on $f: U \to \mathbb{R}^3$ which are linearly independent at each $u \in U$. Then in a neighborhood U_0 of each u_0 we can change variables, $\phi: V_0 \to U_0$, so that $f \circ \phi = \tilde{f}$ has coordinate vector fields $\tilde{f}_{v_i}$ proportional to X_i.*

(This result is *false* for higher dimensional submanifolds of $\mathbb{R}^n$ and, more generally, for any differentiable manifold of dimension >2. See Spivak [A17], vol. I, ch. 6.)

PROOF. 1. Consider the vector fields $\bar{X}_i(u) = df_u^{-1} X_i(u)$ defined for $u \in U$. Suppose we could find a change of variables $\eta: U \to V, \eta(u) = (v^1(u), v^2(u))$, for which

(*) $\qquad dv^1(\bar{X}_2) = 0 \qquad dv^2(\bar{X}_1) = 0.$

Then in terms of canonical basis vector fields $(\tilde{e}_1, \tilde{e}_2)$ on V, $d\eta_u(\bar{X}_1(u)) = dv_u^1(\bar{X}_1(u))\tilde{e}_1 + 0$ and $d\eta_u(\bar{X}_2(u)) = 0 + dv_u^2(\bar{X}_2(u))\tilde{e}_2$. Consequently, if

$\phi = \eta^{-1}: V \to U$, then $\tilde{f} = f \circ \phi$ satisfies $\tilde{f}_{v_i} = d\tilde{f}_v(\tilde{e}_i) = df_{\phi(v)} \circ d\phi_v(\tilde{e}_i) = a_i(v)X_i$, where $a_i(v) = (dv_v^i(\overline{X}_i(\phi(v)))^{-1}$. Thus ϕ is the required change of variables. Note that $a_i(v)$ is well defined, for if $dv_v^i(X_i(\phi(v))) = 0$, then $dv_v^i = 0$ since $\{\overline{X}_1, \overline{X}_2\}$ are linearly independent and $dv^i(\overline{X}_j) = 0$, $i \neq j$. This contradicts the assumption that $\eta = (v^1, v^2)$ is a change of variables.

In order to complete the proof, it is necessary to establish the existence of a pair of functions $v^1(u)$, $v^2(u)$, defined on some neighborhood of u_0, satisfying (*) with $dv^i \neq 0$, $i = 1, 2$. This last condition will ensure that $\eta = (v^1, v^2)$ is a change of variables.

2. Let $\{e_1, e_2\}$ be the canonical basis vector fields on U and write $\overline{X}_i(u) = \sum_{k=1}^{2} \xi_i^k(u)e_k$. By the standard existence theorem for ordinary differential equations, we may assert the existence, locally, of integral curves $c_i(s)$ of $\overline{X}_i(u)$. That is, for $|s|$ sufficiently small, we may find curves $c_1(s)$, $c_2(s)$ in V with $c_i(0) = u_0$ and $\dot{c}_i(s) = \overline{X}_i(c_i(s))$. We wish to solve (*) which is equivalent to

i) $$\frac{\partial v^1}{\partial u^1} \xi_2^1(u) + \frac{\partial v^1}{\partial u^2} \xi_2^2(u) = 0,$$

ii) $$\frac{\partial v^2}{\partial u^1} \xi_1^1(u) + \frac{\partial v^2}{\partial u^2} \xi_1^2(u) = 0,$$

with the initial conditions $v^i(c_i(s)) = s$. A standard result in partial differential equations (see F. John, *Partial Differential Equations*, Springer-Verlag, New York (1971), pp. 15–36) allows us to do this in a neighborhood of u_0, provided that for i) $\dot{c}_1(s)$ and $\overline{X}_2(c_1(s))$ are linearly independent and for ii) $\dot{c}_2(s)$ and $\overline{X}_1(c_2(s))$ are linearly independent. But $\dot{c}_i(s) = \overline{X}_i(c_i(s))$, so these conditions are satisfied by hypothesis. Also, $s = v^i(c_i(s))$, $i \neq j$, implies that

$$1 = \frac{d}{ds}(v^i(c_i(s))) = dv^i(\dot{c}_i(s)) = dv^i(\overline{X}_i(c(s))).$$

Therefore $dv^i \neq 0$, $i = 1, 2$.

Remarks. i) The function $v^1(u)$ (resp. $v^2(u)$) is an *integral* of the differential equation $\dot{c}(s) = \overline{X}_1(c(s))$ (resp. $\dot{c}(s) = \overline{X}_2(c(s))$). An *integral* of a differential equation

(*) $$\dot{x}(s) = f(x(s), s), \qquad x \in U,$$

is a differentiable function $h: U \to \mathbb{R}$ which is nonconstant on any open set and which is constant on integral curves of (*). That is, $h(x(s)) = \text{const}$ or, equivalently, $(d/ds)h(x(s)) = 0$.

ii) If U is simply connected, it is possible to find a globally defined change of variables, $\phi: V \to U$, satisfying the previous theorem. Here is a brief indication of the proof. The theorem gives a way of constructing these coordinates locally near u_0 by mapping (v^1, v^2) into $(c_1(v^1), c_2(v^2))$, where c_1 is the integral curve of X_i beginning at $c_1(v^1)$. This process may be

3.6 Normal Form for a Surface, Special Coordinates

continued to give a regular map from some domain V into U. The only obstruction to getting a diffeomorphism is the possibility that the integral curves of X_i may intersect in two different points. Using simple connectivity, one may show that this is impossible.

3.6.5 Definition. A regular curve $c = f \circ u: I \to \mathbb{R}^3$ on a surface f is called a *line of curvature* if $\dot{c}(t)/|\dot{c}(t)|$ is a principal direction for all $t \in I$.

Remark. Let u_0 be a point where the principal curvatures are different (a nonumbilic point). By the continuity of the principal curvature functions, we can find a neighborhood of u_0 on which $\kappa_1(u) < \kappa_2(u)$. Let $X_1(u)$, $X_2(u)$ denote the associated principal directions. They may be chosen to be differentiable vector fields for the following reason. Since $\kappa_1(u)$, $\kappa_2(u)$ are solutions to $\det(dn_u + \kappa\, df_u) = 0$, they are differentiable. Since $\bar{X}_i(u) = df_u^{-1}X_i(u)$ are solutions to $dn_u \bar{X}_i(u) + \kappa_i(u)\, df_u \bar{X}_i(u) = 0$, they may be chosen to be differentiable. Of course, they are linearly independent. An application of (3.6.4) proves the following lemma.

3.6.6 Lemma. *Let $f: U \to \mathbb{R}^3$ be a surface on which the principal curvatures are not equal at a point u_0. Then there exists a neighborhood U_0 of u_0 and a change of variables $\phi: V_0 \to U_0$ such that the coordinate lines of $\tilde{f} = f \circ \phi$ are lines of curvature.*

Such coordinates are called *principal curvature coordinates*. In principal curvature coordinates, the Weingarten map will have the matrix representation

$$\begin{pmatrix} \kappa_1 g_{11} & 0 \\ 0 & \kappa_2 g_{22} \end{pmatrix}$$

Conversely, using (3.5.5) which shows that this matrix is always equal to $(\sum_k h_{ik} g^{kj})$, Proposition 3.6.7 follows.

3.6.7 Proposition. *If $f: U \to \mathbb{R}^3$ satisfies $h_{12} = g_{12} = 0$, then f is a principal curvature coordinate system.*

We turn our attention now to another naturally occurring vector field on a surface.

3.6.8 Definition. A vector $X \in S_u^1 f \subset T_u f$ is an *asymptotic direction* provided $II_u(X, X) = 0$. The notion of asymptotic direction has invariant geometric meaning.

3.6.9 Proposition. *Asymptotic directions are invariant under isometries and change of variables.*

This proposition is immediate from the properties of II described in (3.3.6). Notice that X is an asymptotic direction if and only if $-X$ is an asymptotic direction. The existence of an asymptotic direction at u is equivalent to the requirement that $\kappa_1 \leq 0$, $\kappa_2 \geq 0$. Therefore an asymptotic direction exists at u if and only if $K(u) \leq 0$ (see 3.6.10 below).

3.6.10 Proposition. i) $K < 0$ if and only if there exists exactly two (up to sign) asymptotic directions.
ii) $K(u) = 0$ and $II_u \neq 0$ if and only if there exists exactly one (up to sign) asymptotic direction.
iii) $K(u) = 0$ and $II_u = 0$ (planar point) if and only if all $X \in S_u^1 f$ are asymptotic directions.

PROOF. i) $K < 0 \Leftrightarrow \det II < 0 \Leftrightarrow II(X, X) = 0$ has precisely two linearly independent solutions, $\pm X$, with $I(X, X) = 1$.
ii) $K = 0$ and $II \neq 0$ means that one of the eigenvalues of II is zero, and the other is equal to $\kappa \neq 0$. If (X_1, X_2) is a basis of eigenvectors with respect to 0, κ, then for any $X = \zeta^1 X_1 + \zeta^2 X_2$, $II(X, X) = \kappa(\zeta^2)^2$. Therefore X_1 is the only principal direction.
iii) Is clear. □

3.6.11 Definition. A regular curve $c = f \circ u: I \to \mathbb{R}^3$ is an *asymptotic line* provided $\dot{c}(t)/|\dot{c}(t)|$ is an asymptotic direction at $u(t)$ for all $t \in I$. The surface $f: U \to \mathbb{R}^3$ is presented in *asymptotic coordinates* near u_0 if the coordinate lines are asymptotic lines in a neighborhood of u_0.

3.6.12 Lemma. *Suppose $K(u_0) < 0$ on $f: U \to \mathbb{R}^3$. Then there is an asymptotic coordinate patch defined on some neighborhood of u_0.*

PROOF. By continuity of the Gauss curvature, there exists a neighborhood of u_0 on which $K < 0$. By (3.6.10, i), there exist two linearly independent asymptotic vector fields X_1, X_2 on some, possibly smaller, simply connected neighborhood of u_0. Now Theorem (3.6.4) completes the proof. □

3.6.13 Examples. 1. For the torus (3.4.5), $K < 0$ on the inside. Thus on the inside there exist precisely two asymptotic directions at each point.
2. On the sphere, no asymptotic directions exist at any point.

Remark. The reason for calling these directions asymptotic becomes clear from the following observation. A regular curve $c(t)$ on the surface has zero normal curvature at $c(t)$, i.e., $II(\dot{c}, \dot{c}) = 0 \Leftrightarrow \ddot{c}(t) \cdot n \circ u(t) = 0 \Leftrightarrow \ddot{c}(t) \in T_u f$. So asymptotic lines have no normal component of acceleration. In particular, if $c(t)$ is a straight line in $\mathbb{R}^3$ which lies on the surface, $\ddot{c}(t) = 0$ and c is an asymptotic curve.

3.7 Special Surfaces; Developable Surfaces

3.7.1 Definition. A *triply orthogonal system of surfaces* is a differentiable map $F: W \to \mathbb{R}^3$, defined on an open set $W \subset \mathbb{R}^3$, satisfying:

i) $dF_{(u,v,w)}: T_{(u,v,w)}\mathbb{R}^3 \to T_{F(u,v,w)}\mathbb{R}^3$ is bijective for all $(u, v, w) \in W$.
ii) $F_u \cdot F_v = F_u \cdot F_w = F_v \cdot F_w = 0$.

3.7 Special Surfaces; Developable Surfaces

Remark. The reason for calling such a map by this extraordinary name is that at each $p = (u_0, v_0, w_0) \in W$, the three surfaces

$$(u, v) \mapsto F(u, v, w_0)$$
$$(v, w) \mapsto F(u_0, v, w)$$
$$(u, w) \mapsto F(u, v_0, w)$$

are mutually orthogonal. We will denote these surfaces by f^{w_0}, f^{u_0}, and f^{v_0}, respectively. They are regular by (i).

Notice that by condition (ii), not only are the surfaces orthogonal, but $g_{12} = 0$ on each of them. Furthermore, $F_w(u, v, w_0)$ is normal to f^{w_0} at (u, v, w_0) (and the identical relation holds for the other two surfaces) and, differentiating,

$$(F_u \cdot F_v)_w = (F_u \cdot F_w)_v = (F_v \cdot F_w)_u = 0.$$

Therefore $F_{uv} \cdot F_w = F_{uw} \cdot F_v = F_{vw} \cdot F_u = 0$, which means that $h_{12} = 0$ on each of the surfaces. By (3.6.7), we may conclude that

3.7.2 Proposition (Dupin). *The coordinate curves on a surface in a triply orthogonal system are lines of curvature.*

3.7.3 An example. *Second order confocal surfaces.* Let $0 < c < b < a$ and consider the equation

$$\psi(\rho) = \frac{x^2}{c - \rho} + \frac{y^2}{b - \rho} + \frac{z^2}{a - \rho} - 1 = 0.$$

For $\begin{cases} \rho < c \\ c < \rho < b \\ b < \rho < a \end{cases}$

the solution set of this equation is $\begin{cases} \text{an ellipsoid} \\ \text{a hyperboloid of one sheet} \\ \text{a hyperboloid of two sheets.} \end{cases}$

Let $Q = \{(x, y, z) \in \mathbb{R}^3 \mid x > 0, y > 0, z > 0\}$ be the positive quadrant. Let $W =]-\infty, c[\times]c, b[\times]b, a[\subset \mathbb{R}^3$.

Now we observe that for each $(x, y, z) \in Q$ there exists a unique triple $(u, v, w) \in W$ such that if $\rho = u$, (x, y, z) lies on an ellipsoid, if $\rho = v$, (x, y, z) lies on a hyperboloid of one sheet, and if $\rho = w$, (x, y, z) lies on a hyperboloid of two sheets. To see this we simply consider the equations $\psi(u) = \psi(v) = \psi(w) = 0$ and solve for x, y, and z:

$$x^2(u, v, w) = (c - u)(c - v)(c - w)/(c - b)(c - a)$$
$$y^2(u, v, w) = (b - u)(b - v)(b - w)/(b - a)(b - c)$$
$$z^2(u, v, w) = (a - u)(a - v)(a - w)/(a - b)(a - c).$$

3 Surfaces: Local Theory

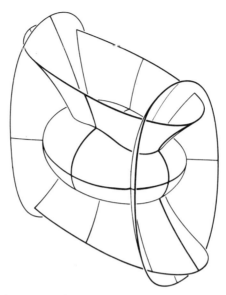

Figure 3.8   Confocal surfaces, second order

These formulae express x, y, and z uniquely as functions of (u, v, w) in the required domain W. Remember, x, y, and z are assumed to be strictly positive.

Now consider the map $F: W \to Q$ given by

$$(u, v, w) \mapsto (x(u, v, w), y(u, v, w), z(u, v, w)).$$

We claim that this is a triply orthogonal system. To see this we shall take a geometric approach and show that the surfaces $\psi(u)$, $\psi(v)$, and $\psi(w)$ are regular and mutually orthogonal. Since

$$\operatorname{grad} \psi(v) = 2(x/(c - v), y/(b - v), z/(a - v)) \neq (0, 0, 0)$$
$$\operatorname{grad} \psi(w) = 2(x/(c - w), y/(b - w), z/(a - w)) \neq (0, 0, 0)$$
$$\operatorname{grad} \psi(u) = 2(x/(c - u), y/(b - u), z/(a - u)) \neq (0, 0, 0),$$

we conclude that, for example,

$\operatorname{grad} \psi(v) \cdot \operatorname{grad} \psi(w)$

$$= 4(x^2/(c - v)(c - w) + y^2/(b - v)(b - w) + z^2/(a - v)(a - w)) = 0$$
$$\operatorname{grad} \psi(v)^2 = 4(u - v)(w - v)/(a - v)(b - v)(c - v)$$
$$\operatorname{grad} \psi(w)^2 = 4(u - w)(v - w)/(a - w)(b - w)(c - w).$$

Here we use the above equations for x^2, y^2, z^2. Regularity and orthogonality are established by these formulae.

3.7.4 Definition. A surface $f: U \to \mathbb{R}^3$ is a *ruled surface* if every $u_0 \in U$ has a neighborhood on which we may define a change of variables $u = \phi(s, t)$ so that

$$\tilde{f}(s, t) = f \circ \phi(s, t) = sX(t) + c(t).$$

3.7 Special Surfaces; Developable Surfaces

Here $X(t)$ is a vector field along a curve $c(t)$ on f. The curves $t = $ constant are lines in $\mathbb{R}^3$ and are called *generators* of f. A curve $s = $ constant is called a *directrix*.

If, in addition, the normal vector field $\tilde{n}(s, t)$ is a constant along generators, i.e., $\tilde{n}_s = 0$, then f is called *developable*.

3.7.5 Proposition. i) *On a ruled surface, generators are asymptotic curves. Consequently, $K \leq 0$.*
ii) *A ruled surface, f, is developable $\Leftrightarrow$*
In (s, t)-parameters f_{st} is a linear combination of f_s and f_t $\Leftrightarrow$
$K = 0$ on f.

PROOF. i) In (s, t)-parameters, $f_{ss} = 0$. Therefore $h_{11} = II(f_s, f_s) = -n_s \cdot f_s = n \cdot f_{ss} = 0$, and so $K = -h_{12}^2/\det(g_{ij}) \leq 0$.
ii) In (s, t)-parameters, we have shown in (i) that $n_s \cdot f_s = 0$. Therefore if $f(s, t)$ is a ruled surface, $n_s = 0 \Leftrightarrow n_s \cdot f_t = n_s \cdot f_s = 0 \Leftrightarrow n_s \cdot f_t = 0 \Leftrightarrow n \cdot f_{st} = 0 \Leftrightarrow h_{12} = 0 \Leftrightarrow K = -h_{12}^2/\det(g_{ij}) = 0$. □

3.7.6 Examples of developable surfaces

1. *Tangential developables.* Consider a space curve $c: I \to \mathbb{R}^3$ with $\dot{c}(t), \ddot{c}(t)$ linearly independent for all t. The surface $f(s, t) = s\dot{c}(t) + c(t)$, $s \neq 0$, is called the *tangential developable of c*. Since $f_{st} = \ddot{c}(t)$, it is a linear combination of $f_s = \dot{c}(t)$ and $f_t = s\ddot{c}(t) + \dot{c}(t)$.

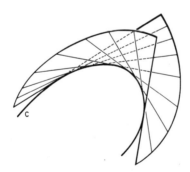

Figure 3.9 Tangential developable

2. *Cylinder over a curve.* Let $c(t)$ be a plane curve and $X_0 \neq 0$ a vector not lying in the plane of the curve. The surface $f(s, t) = sX_0 + c(t)$, a generalized cylinder, is a developable surface.

3. *Cone.* The surface $f(s, t) = sX(t) + x$, $s \neq 0$, $X(t)$ and $\dot{X}(t)$ linearly independent, is a *cone with vertex x*. It is easily seen that f is a developable surface.

Developable surfaces enter into the general theory of surfaces via the following construction.

3 Surfaces: Local Theory

3.7.7 Proposition (Existence of an *osculating developable*). *Let $c(t) = f \circ u(t)$ be a regular curve on a surface f. Suppose $Y(t)$ is a vector field along $c(t)$ tangential to f, satisfying $II(\dot{c}(t), Y(t)) = 0$, and linearly independent of $\dot{c}(t)$. Then $g(s, t) = sY(t) + c(t)$ is a developable surface.*

PROOF. Easy: $g_{st} = \dot{Y}$, $n \cdot g_{st} = n \cdot \dot{Y} = -\dot{n} \cdot Y = II(\dot{c}, Y) = 0$. □

Note that the surface f and the constructed developable surface g both contain the curve c, and at each point of c they have identical tangent planes. We will exploit these facts in a very important geometric construction (parallel translation) in (4.2.5) and (4.4.3). For the time being, we will be content to carry out the construction explicitly in a simple case.

3.7.8 Example. On the sphere

$$f(u, v) = (\cos u \cos v, \cos u \sin v, \sin u), \quad (u, v) \in \,]-\pi/2, \pi/2[\, \times \mathbb{R},$$

consider the latitude circle $c(t) = f(u(t), v(t))$, $u(t) = a$, $v(t) = t/\cos a$, $a \in \,]-\pi/2, \pi/2[\,$, $t/\cos a \in \,]-\pi, \pi[\,$. It follows that $\dot{c}(t) = f_v/\cos a$, so $|\dot{c}(t)| = 1$. Let $Y(t) = f_u(a, t/\cos a) = (-\sin a \cos v(t), -\sin a \cos v(t), \cos a)$. In (3.3.7) we showed that $II = -I$ and that $f_u \cdot f_v = 0$. Consequently, $II(\dot{c}(t), Y(t)) = 0$, which means $g(s, t) = sY(t) + c(t)$ is an osculating developable surface.

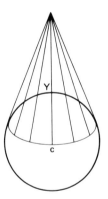

Figure 3.10 Osculating cone

In the case $a \neq 0$, $g(s, t)$ is a circular cone all of whose generators pass through the point $(0, 0, 1/\sin a)$ when $s = \cot a$. In the case $a = 0$, $c(t)$ is the equator and $g(s, t)$ is a right circular cylinder.

We finish this section by looking more closely at surfaces with Gauss curvature equal to zero. We have already shown in (3.7.5) that, in the class of ruled surfaces, developable surfaces are precisely those with $K = 0$. The question remains: are there surfaces with $K = 0$ which are *not* ruled and hence not developable?

The answer is given locally by the following theorem.

58

3.7 Special Surfaces; Developable Surfaces

3.7.9 Theorem. *A surface $f: U \to \mathbb{R}^3$ without planar points is developable if and only if $K = 0$.*

Remark. Recall that f is planar at $u_0 \in U$ if $II_{u_0} \equiv 0$. The theorem fails to be true if the hypothesis of "no planar points" is dropped. An explicit counterexample is constructed in (3.9.4).

PROOF. If f is developable, we know by (3.7.5) that $K = 0$. Conversely, let $K = 0$. The absence of planar points allows us to assert the existence of unique (up to sign) mutually orthogonal principal curvature vector fields in a neighborhood of each point u_0 (see (3.5.3)). Using (3.6.6), we may introduce new coordinates (v^1, v^2) on a neighborhood U_0 of u_0 such that the v^1-coordinate curves are integral curves of the principal curvature vector field corresponding to the principal curvature $\kappa_1 = 0$. Without loss of generality, we may assume that $(0, 0) \mapsto u_0$.

We change variables once more. Let $(s, t) \to (v^1(s), t)$, where $v^1(s)$ is the inverse of the arc-length function along the curve $f(v^1, 0)$. Clearly, $\partial v^1(s)/\partial s \neq 0$. Therefore $\tilde{f} = f \circ \phi$ is a new coordinatization defined in a neighborhood of $(0, 0)$ with $\phi(0, 0) = 0$. In this new coordinatization, both $\tilde{f}(s, 0)$ and $\tilde{f}(0, t)$ are parameterized by arc length. The vector $\tilde{f}_s(0, 0)$ is a principal direction corresponding to $\kappa_1 = 0$.

Agreeing to write $f(s, t)$ instead of $\tilde{f}(s, t)$, let us show that $f_{ss} = 0$. First observe that $n_s = -\kappa_1 f_s = 0$, $n_t = -\kappa_2 f_t \neq 0$, $f_s \cdot f_t = 0$, and $f_s \cdot n_t = 0$. This implies that $f_{ss} \cdot n = -f_s \cdot n_s = 0$, and therefore f_{ss} is purely tangential. Now

$$f_{ss} \cdot f_t = \left(-\frac{1}{\kappa_2}\right) f_{ss} \cdot n_t = \left(\frac{1}{\kappa_2}\right) f_s \cdot n_{st} = 0,$$

so f_{ss} is a multiple of f_s. But $f_s^2(s, 0) = 1$ and $f_s^2(s, t)_{,t} = 2 f_s \cdot f_{st} = -2 f_{ss} \cdot f_t = 0$, which implies $f_s^2(s, t) = 1$. Differentiating this equation, we see $f_{ss} \cdot f_s = 0$. Therefore $f_{ss} = 0$.

This means that the s-parameter curves are straight lines, parameterized by arc length.

Letting $c(t) = f(0, t)$, we see that $f(s, t) = sX(t) + c(t)$, where $X(t) = f_s(0, t)$. Thus $f(s, t)$ is a ruled surface with $n_s = 0$, i.e., a developable surface. □

Remark. Even though we have shown that flat surfaces without planar points are developable surfaces, we still have not completely described how a piece of surface with $K = 0$ can look in $\mathbb{R}^3$. Even without admitting planar points, one can patch together developable surfaces in a variety of ways, cf. Figure 3.11.

The following proposition shows that developable surfaces look basically like those described in (3.7.6)

3.7.10 Proposition. *Suppose $f: U \to \mathbb{R}^3$ is a developable surface without planar points. Then on an open dense set $A \subset U$, f is either a cylinder, a cone, or a tangential developable.*

3 Surfaces: Local Theory

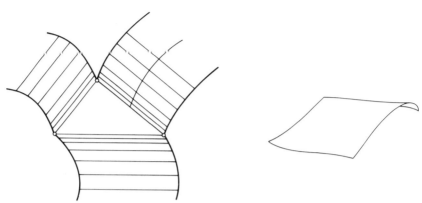

Figure 3.11 Some flat surfaces (Adapted from Manfredo P. do Carmo, *Differential Geometry of Curves and Surfaces*, Prentice-Hall, Inc., 1976, p. 409.)

PROOF. 1. By (3.7.9) we may assume that f can be written locally as $f(s, t) = sX(t) + c(t)$ for (s, t) within some neighborhood $U = I \times J$ of $(0, 0)$, with $f_s \cdot f_t = 0, n_s \cdot f_t = n \cdot f_{st} = 0$. Therefore $X \cdot (s\dot{X} + \dot{c}) = 0, \dot{c}(t) = f_t(0, t) \neq 0$, $f_s = X \neq 0$, and $n \cdot \dot{X} = 0$. The tangent space $T_{(s,t)}f$ is spanned by $X(t)$ and $\dot{c}(t)$, $X(t) \cdot \dot{c}(t) = 0$. Since $\dot{X}(t) \in T_{(s,t)}f$ and $X \cdot \dot{X} = 0$, $\dot{X}(t) = r(t)\dot{c}(t)$ for some real-valued differentiable function $r(t)$.

2. Let $t \in I$. Let I_0 be the set of $t \in \mathbb{R}$ satisfying one of the following properties.
   a) There exists a neighborhood $U(t_0)$ of t_0 on which $r(t) = 0$.
   b) There exists a neighborhood $U(t_0)$ of t_0 on which $r(t) = $ constant $\neq 0$.
   c) There exists a neighborhood $U(t_0)$ of t_0 on which $r(t) \neq 0$ and $\dot{r}(t) \neq 0$.
   By definition, $I_0 \subset I$ is open. A moment's reflection will show that I_0 is also a dense subset of I. In fact I_0 is the union of the sets where $r(t) \neq 0$ and $\dot{r}(t) \neq 0$ with the interior points of the set where $\dot{r}(t) = 0$. We will now show that the cases (a), (b), and (c) correspond to a cylinder, a cone, and a tangential developable, respectively.

3. Suppose $t_0 \in I_0$ satisfies (a). Then $X(t) = X_0 = $ constant, so $f(s, t) = sX_0 + c(t)$, a cylinder. Suppose $t_0 \in I_0$ satisfies (b). Then $X(t) - X(t_0) = r_0(c(t) - c(t_0))$. Therefore $f(s, t) = ((s + 1)/r_0)X(t) + (c(t_0) - X(t_0)/r_0)$, a cone with vertex $(c(t_0) - X(t_0)/r_0)$. Suppose $t_0 \in I_0$ satisfies (c). Let $\tilde{c}(t) = -X(t)/r(t) + c(t)$. Then $\dot{\tilde{c}} = \dot{r}X/r^2$, so $\dot{\tilde{c}}$ is linearly independent from $\tilde{c}$ since $\dot{X}$ and X are orthogonal. We may write $c(t) = \tilde{c}(t) + X(t)/r(t) = \tilde{c}(t) + r(t)\dot{\tilde{c}}(t)/\dot{r}(t)$. If we let $\tilde{s} = \tilde{s}(s, t) = sr^2(t)/\dot{r}(t) + r(t)/\dot{r}(t)$, we may write $f(s, t) = \tilde{f}(\tilde{s}, t) = \tilde{s}\dot{\tilde{c}}(t) + \tilde{c}(t)$, a tangential developable. □

Remark. It is still not clear from (3.7.10) whether, for example, the local coordinates expressing f as a cone, cylinder, or tangential developable, can be extended *along the generators* (i.e., in the s-direction) to the boundary of f.

There is a strong *global* result concerning surfaces with $K \equiv 0$. If $f: U \to \mathbb{R}^3$ is assumed to be *geodesically complete* (see (6.4.4) for the definition), then any surface with $K \equiv 0$ must be a generalized cylinder.[2] This result was first proved by Pogorelov.[3] Note that it is not necessary to assume that f has no planar points.

3.8 The Gauss and Codazzi–Mainardi Equations

Before we begin this section let us agree to abbreviate our notation for partial derivatives. We will write ϕ_i, or occasionally $\phi_{,i}$, for $\partial\phi/\partial u^i = \phi_{u^i}$. When higher order partial derivatives occur, we will treat them in the same fashion, writing ϕ_{ik} for $\phi_{u^i u^k}$, etc. The matrices $(g_{ik}(u))$ and $(h_{ik}(u))$ will denote matrix representations of the first and second fundamental forms with respect to the standard basis $\{e_i\}$ of $T_u \mathbb{R}^2$ and $\{f_i\}$ of $T_u f$. The inverse of (g_{ij}) will be denoted by (g^{ij}).

3.8.1 Theorem

$$(*) \quad f_{ik}(u) = \sum_l \Gamma^l_{ik}(u) f_l(u) + h_{ik}(u) n(u), \quad n_i(u) = -\sum_{l,k} h_{il} g^{lk} f_k(u),$$

where

$$(**) \quad \Gamma^l_{ik} := \sum_j g^{lj} f_{ik} \cdot f_j = \frac{1}{2} \sum_j g^{lj} (g_{ij,k} + g_{jk,i} - g_{ki,j}),$$

where $g_{ij,k} = \partial g_{ij}/\partial u^k$.

3.8.2 Definition. The six functions $\Gamma^l_{ik}(u) = \Gamma^l_{ki}(u)$ in (*) are called the *Christoffel symbols of the second kind*. The functions

$$\Gamma_{kij} = \tfrac{1}{2}(g_{ij,k} + g_{jk,i} - g_{ki,j})$$

are called the *Christoffel symbols of the first kind*.

Remark. The expressions (*) and (**) express f_{ik} and n_i in terms of the Gauss frame (f_1, f_2, n). Moreover, the coefficients can be expressed in terms of the g_{ik}, h_{ik}, and $g_{ik,l}$.

PROOF. 1. Since $(f_1(u), f_2(u), n(u))$ span $T_{f(u)} \mathbb{R}^3$, we may write $f_{ik} = f_{ki} = \sum_l \Gamma^l_{ik} f_l + a_{ik} n$, where the coefficients are to be determined. By taking the

---

[2] A *generalized cylinder* in $\mathbb{R}^3$ is a surface, S, that may be described as follows: there exists a curve $c(t)$, $c: \mathbb{R} \to \mathbb{R}^3$, and a fixed direction n such that $f(s, t) = c(t) + sn$, $f: \mathbb{R}^2 \to \mathbb{R}^3$ is a global parameterization of S.

[3] Pogorelov, A. W. Extension of the theorem of Gauss on the spherical image of surfaces of bounded extrinsic curvature. *Dokl. Akad. Nauk*, **111**, 945–947 (1956) (Russian).
   Other proofs of this theorem were given by P. Hartman and L. Nirenberg (1959) and J. J. Stoker (1969). A quite simple proof with a list of references on the topic may be found in Massey, W. S. Surfaces of Gaussian curvature zero in Euclidean 3-space. *Tôhoku Math. J.* (2), **14**, 73–74 (1962).

3 Surfaces: Local Theory

inner product with n, we see that $h_{ik} = a_{ik}$. By taking the inner product with f_j, we get

$$f_{ik} \cdot f_j = \sum_l \Gamma^l_{ik} g_{lj},$$

and therefore

$$\Gamma^l_{ik} = \sum_l g^{lj} f_{ik} \cdot f_j = \Gamma^l_{ki}.$$

Furthermore,

(α) $\qquad g_{ij,k} = (f_i \cdot f_j)_k = f_{ik} \cdot f_j + f_i \cdot f_{jk} = \sum_l \Gamma^l_{ik} g_{lj} + \sum_l \Gamma^l_{jk} g_{li},$

and cyclical permutation of the indices yields:

(β) $\qquad g_{ki,j} = \sum_l \Gamma^l_{kj} g_{li} + \sum_l \Gamma^l_{ij} g_{lk}$

(γ) $\qquad g_{jk,i} = \sum_l \Gamma^l_{ji} g_{lk} + \sum_l \Gamma^l_{ki} g_{lj}.$

The equations in (**) are equivalent to (α) $-$ (β) $+$ (γ).
2. The expression for $n_i = n_{u^i}$ follows from (3.5.5). $\qquad \square$

3.8.3 Theorem (Integrability conditions). *The equations $f_{ijk} = f_{ikj}$ and $n_{ij} = n_{ji}$ are equivalent to the following relations between g_{ik}, h_{ik}, $g_{ik,l}$, $h_{ik,l}$, and $\Gamma^k_{ij,l}$.*

i) $\qquad \Gamma^m_{ij,k} - \Gamma^m_{ik,j} + \sum_l (\Gamma^l_{ij} \Gamma^m_{lk} - \Gamma^l_{ik} \Gamma^m_{lj}) = \sum_l (h_{ij} h_{kl} - h_{ik} h_{jl}) g^{lm}.$

ii) $\qquad \sum_l \Gamma^l_{ij} h_{lk} - \sum_l \Gamma^l_{ik} h_{lj} + h_{ij,k} - h_{ik,j} = 0.$

The equations (i) are called the *Gauss equations*, and the equations (ii) the *Codazzi–Mainardi equations*.

Remark. The Gauss equations come from equating of the coefficients of f_m in the equations $f_{ijk} = f_{ikj}$. The Codazzi–Mainardi equations came from equating the coefficients of n in the equations $f_{ijk} = f_{ikj}$. Equating the coefficients of f_m and n in $n_{ij} = n_{ji}$ gives another derivation of the Codazzi–Mainardi equations.

PROOF. 1. Let $f_{ijk} = \sum_m A^m_{ijk} f_m + B_{ijk} n$. Using (3.8.1) (*), we may express A^m_{ijk} as

$$A^m_{ijk} = \Gamma^m_{ij,k} + \sum_l \Gamma^l_{ij} \Gamma^m_{lk} - \sum_l h_{ij} h_{kl} g^{lm}.$$

Since $A^m_{ijk} = A^m_{ikj}$, interchanging j and k and subtracting proves (i).
2. Another application of (3.8.1) (*) enables us to write

$$B_{ijk} = \sum_l \Gamma^l_{ij} h_{lk} + h_{ij,k}.$$

3.8 The Gauss and Codazzi–Mainardi Equations

Since $B_{ijk} = B_{ikj}$, this proves (ii).

3. Let

$$C_{ij}^k = -\left(\sum_l h_{il} g^{lk}\right)_{,j} - \sum_{l,m} h_{il} g^{lm} \Gamma^k_{mj}.$$

Using (**) to obtain an expression for Γ^k_{mj} and also the fact that

$$\sum_k g^{mk}_{,j} g_{ki} = -\sum_k g^{mk} g_{ki,j}$$

(obtained by differentiating $\sum_k g^{mk} g_{ki} = \delta^m_i$) enables us to conclude that $C_{ij}^k = C_{ji}^k$. This is equivalent to (ii). □

3.8.4 Definition. The *curvature tensor* of f is the collection of functions $R_{iljk} = \sum_m g_{lm} R^m_{ijk}$, $1 \le i, j, k, l \le 2$, where

$$R^m_{ijk} := \Gamma^m_{ij,k} - \Gamma^m_{ik,j} + \sum_l (\Gamma^l_{ij} \Gamma^m_{lk} - \Gamma^l_{ik} \Gamma^m_{lj}), \qquad 1 \le i, j, k, m \le 2.$$

3.8.5 Lemma

$$R_{iljk} = h_{ij} h_{kl} - h_{ik} h_{jl}.$$

Consequently, $R_{iljk} = -R_{ilkj} = -R_{lijk} = R_{jkil}$ *and the curvature tensor is totally determined by* $R_{1212} = -R_{2112} = R_{2121} = -R_{1221} = \det(h_{ij})$; *all the other* R_{iljk} *are equal to zero.*

PROOF. An immediate consequence of (3.8.3, i). □

The following theorem will show that the curvature tensor has a geometric meaning in the sense that it is the coordinate expression of a multilinear map from $T_u f \times T_u f \times T_u f \times T_u f$ into $\mathbb{R}$ which is *independent of the choice of coordinates*. In contrast to this, the Christoffel symbols Γ^k_{ij} are *not* coordinate independent.

3.8.6 Theorem. *Let* $f: U \to \mathbb{R}^3$ *be a surface. Let*

$$X = \sum_i \xi^i f_i, \qquad Y = \sum_i \eta^i f_i, \qquad Z = \sum_j \zeta^j f_j, \qquad W = \sum_k \omega^k f_k,$$

be four tangential vector fields. Then the multilinear form

$$R: T_u f \times T_u f \times T_u f \times T_u f \to \mathbb{R}$$

given by $R(X, Y, Z, W) = \sum_{i,l,j,k} R_{iljk} \xi^i \eta^l \zeta^j \omega^k$ *has the following properties:*
i) $R(X, Y, Z, W) = -R(Y, X, Z, W) = R(Y, X, W, Z) = R(Z, W, X, Y)$
   $R(f_i, f_l, f_j, f_k) = R_{iljk}.$
ii) R *is linear in each variable.*
iii) *Let* $\phi: V \to U$ *be a change of variables and* $\tilde{R}_{iljk}$ *be the curvature tensor associated to* $\tilde{f} = f \circ \phi$. *Then* $\tilde{R}(X, Y, Z, W) = R(X, Y, Z, W)$.

3 Surfaces: Local Theory

PROOF. 1. (i) and (ii) follow directly from (3.8.5) and the definition of R.
2. Writing $\phi(v) = (u^1(v^1, v^2), u^2(v^1, v^2))$, we may write $\tilde{R}$ in terms of R as follows:

(*) $\qquad \tilde{R}_{iljk} = \sum_{i'l'j'k'} R_{i'l'j'k'} u_i^{i'} u_l^{l'} u_j^{j'} u_k^{k'}$, where $u_i^{i'} = \partial u^{i'}/\partial v^i$, etc.

This expression may be derived from (3.8.5) by plugging in the expression of $\tilde{h}_{ij}$ in terms of h_{ij}, (3.3.6):

$$\tilde{h}_{ij} = \pm \sum_{i'j'} h_{i'j'} u_i^{i'} u_j^{j'}.$$

If $X = \sum_i \xi^{i'} f_{i'} = \sum_i \tilde{\xi}^i \tilde{f}_i = \sum_{i,i'} \tilde{\xi}^i u_i^{i'} f_{i'}$, etc., then (*) implies that

$$\sum_{i,l,j,k} \tilde{R}_{iljk} \tilde{\xi}^i \tilde{\eta}^l \tilde{\zeta}^j \tilde{\omega}^k = \sum_{i',l',j',k'} R_{i'l'j'k'} \xi^{i'} \eta^{l'} \zeta^{j'} \omega^{k'}. \qquad \square$$

3.8.7 Theorema Egregium (Gauss).[4] *The Gauss curvature $K(u)$ can be computed from the first fundamental form and its first and second partial derivatives. More precisely,*

$$K(u) = \frac{R_{1212}(u)}{\det(g_{ik}(u))}.$$

PROOF. R_{1212} is defined in terms of (g_{ij}) and its first and second partials by (3.8.4). The formula for K is (3.5.5). Now use (3.8.5). $\qquad \square$

The meaning of this "celebrated theorem" of Gauss will be examined in the next chapters where we will explore the intrinsic theory of surfaces. Suffice it to say now that Gauss curvature, defined in terms of the second fundamental form (which is dependent on how the surface sits in space), can be computed from a knowledge of the first fundamental form and its partial derivatives. The latter quantities can be computed, in principle, by a resident of the surface, without knowledge of or reference to the shape of the surface in $\mathbb{R}^3$.

To end this chapter, we will prove an analogue of the existence and uniqueness theorem for curves in $\mathbb{R}^n$, (1.3.5) and (1.3.6).

3.8.8 Theorem (Fundamental theorem of surface theory). *Let U be an open, simply-connected subset of $\mathbb{R}^2$. Suppose I_u, II_u are quadratic forms on $T_u\mathbb{R}^2$, $u \in U$, whose coefficients $(g_{ik}(u))$ and $(h_{ik}(u))$ are differentiable functions of u. If I_u is positive definite and the Gauss and Codazzi–Mainardi equations (3.8.3) are satisfied, then:*

i) *There exists a surface $f: U \to \mathbb{R}^3$ whose first and second fundamental forms are I_u and II_u.*

---

[4] Gauss, C. F. Disquisitiones generales circas superficies curvas. *Commentationes societatis regiae scientiarum Gottingensis recentiores*, **6**, Göttingen, 1828.

3.8 The Gauss and Codazzi–Mainardi Equations

ii) *Any two surfaces f and $\tilde{f}$ defined on U which have the same first and second fundamental form differ by an isometry:*

$$\tilde{f} = B \circ f, \quad B \text{ an isometry of } \mathbb{R}^3.$$

PROOF. 1. *The existence of f.* The structural equations of (3.8.1) may be considered as a system of linear partial differential equations for the three $\mathbb{R}^3$-valued functions $f_1(u), f_2(u), n(u)$. The integrability conditions $f_{i,jk} = f_{i,kj}$, $n_{i,j} = n_{j,i}$ are satisfied (this is the content of the Gauss and Codazzi equations). By a well-known theorem of differential equations (see Flanders [B8], pp. 92–101, or Spivak [A15], Vol. I, ch. 6), there exists a unique solution to this system satisfying any given initial conditions $f_i(u_0) = X_i$, $n(u_0) = N$, where $X_i \cdot X_k = g_{ik}(u_0)$, $X_i \cdot N = 0$, $|N| = 1$, and (X_1, X_2, N) is positively oriented.

Choose $x_0 \in \mathbb{R}^3$, and let

$$f(u) = \int_{u_0}^{u} \sum_i f_i(u)\, du^i + x_0.$$

Since $f_{1,2} = f_{2,1}$, this integral is independent of path and therefore $f(u)$ is well defined. We wish to show that f is the desired surface. Toward that end, consider the functions $f_i \cdot f_j(u)$, $n \cdot f_j(u)$, $n \cdot n(u)$. Because f_i and n satisfy the differential equations (3.8.1), we have

$$(f_i \cdot f_j)_{,k} = \sum_l \Gamma_{ik}^l (f_l \cdot f_j) + \sum_l \Gamma_{jk}^l (f_i \cdot f_l) + h_{ik}(n \cdot f_j) + h_{jk}(n \cdot f_i),$$

$$(n \cdot f_j)_{,i} = -\sum_{k,l} h_{il} g^{lk}(f_k \cdot f_j) + \sum_l \Gamma_{ij}^l (f_l \cdot n) + h_{ij}(n \cdot n),$$

$$(n \cdot n)_{,i} = -2 \sum_l h_{il} g^{lk}(f_k \cdot n).$$

It is easily seen that these differential equations would be satisfied if $f_i \cdot f_j = g_{ij}$, $n \cdot f_j = 0$, $n \cdot n = 1$. Our functions agree with these functions at $u = u_0$, and therefore must be equal to these functions on U. From $f_i \cdot f_j = g_{ij}$ we may conclude that f_1, f_2 are linearly independent, which implies that f is indeed a surface. Furthermore, $\det(f_1, f_2, n) > 0$ when $u = u_0$, and since it never equals zero, it must be positive everywhere on U. The second fundamental form of f is determined by $-n_i \cdot f_k$. Using the differential equations (*) of (3.8.1) for which n and f_k are solutions, we see that $-n_i \cdot f_k = h_{ik}$. Therefore f is the desired surface.

2. *Uniqueness of f up to isometry.* Suppose f and $\tilde{f}$ are two solutions determined by the initial conditions x_0, X_1, X_2, N and $\tilde{x}_0, \tilde{X}_1, \tilde{X}_2, \tilde{N}$, respectively. Since $X_i \cdot X_k = \tilde{X}_i \cdot \tilde{X}_k$, $X_i \cdot N = \tilde{X}_i \cdot \tilde{N} = 0$, $N \cdot N = \tilde{N} \cdot \tilde{N} = 1$, there exists a unique isometry B such that

$$Bx_0 = \tilde{x}_0, \quad dB_{x_0} X_i = \tilde{X}_i, \quad dB_{x_0} N = \tilde{N}.$$

Since both (X_1, X_2, N) and $(\tilde{X}_1, \tilde{X}_2, \tilde{N})$ are positively oriented, B is a congruence.

3 Surfaces: Local Theory

Since dBf_i, dBn and $\tilde{f}_i$, $\tilde{n}$ satisfy the same system of differential equations with the same initial conditions at $u = u_0$, it follows that $dBf_i = \tilde{f}_i$. Therefore $Bf(u) = Bf(u_0) + \tilde{f}(u) - \tilde{f}(u_0) = \tilde{f}(u)$. □

3.9 Exercises and Some Further Results

3.9.1 Surfaces of revolution with constant Gauss curvature. Consider a surface of revolution given as in (3.3.7) by

$$f(u, v) = (h(u) \cos v, h(u) \sin v, k(u)).$$

Assume that $h'^2 + k'^2 = 1$ and hence that $k'k'' = -h''h'$.
Prove: $g_{11} = 1, g_{12} = 0, g_{22} = h^2, h_{11} = -k'h'' + h'k'', h_{12} = 0, h_{22} = hk'$.
Therefore

$$K = \frac{(h'k'k'' - k'k'h'')}{h} = -\frac{h''}{h}.$$

The requirement that f have constant Gauss curvature K_0 means that h must satisfy

$$h''(u) + K_0 h(u) = 0.$$

Conversely, a function $h(u)$ satisfying this equation with $h'^2 \le 1$ will enable us to construct a surface of constant Gauss curvature K_0.
  Case (i). $K_0 = 0$. Without loss of generality, $h(u) = au + b, 0 \le a \le 1$. If $a = 0$, the surface of revolution is a right-circular cylinder. If $0 < a < 1$, the surface is a circular cone. If $a = 1$, the generated surface is a piece of a plane.

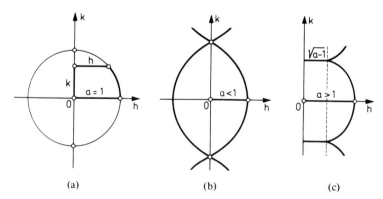

Figure 3.12 (a) Sphere; (b) spindle; (c) bead

  Case (ii). $K_0 = 1$. Without loss of generality, $h(u) = a \cos u$, where $a > 0$ and $a^2 \sin^2 u \le 1$. This implies that $k(u) = \int_0^u \sqrt{1 - a^2 \sin^2 t} \, dt$. When $a = 1$ we get a sphere, when $0 < a < 1$ a spindle-like surface, and when $a > 1$ the surface looks like a column of water about to break into beads.

Case (iii). $K_0 = -1$. Then we may write $h(u) = ae^u + be^{-u}$, requiring $(ae^u - be^{-u})^2 \leq 1$. Consider the case where $b = 1$, $a = 0$. Then

$$h(u) = e^{-u}, \qquad k(u) = \int_0^u \sqrt{1 - e^{-2t}}\, dt, \qquad u \geq 0.$$

The curve $(h(u), k(u))$ in the (x, z) plane is the *tractrix*. It is characterized by the fact that distance, along the tangent line to $(h(u), k(u))$, from $(h(u), k(u))$ to the z-axis is always equal to 1. The surface of revolution is called the *pseudosphere*. It was an important example in the early history of non-Euclidean geometry.

If $ab \neq 0$, then it can be shown that $a = -b = c/2$ or $a = b = c/2$. In the first case the surface of rotation looks something like cones stacked point to point and base to base. In the second case, the surface looks like a horizontally fluted column (see Figure 3.13).

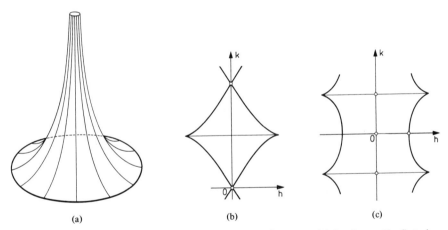

Figure 3.13 (a) The pseudosphere; (b) a pile of cones; (c) horizontally-fluted column

3.9.2 Caustic surfaces.[5] Suppose $f: U \to \mathbb{R}^3$ is a surface whose principal curvatures κ_1 and κ_2 are nonzero and unequal. Let (u^1, u^2) be principal curvature coordinates.

Prove: The functions $b_i(u) = f(u) + n(u)/\kappa_i(u)$, $i = 1, 2$, are surfaces if and only if $\kappa_{1,1}\kappa_{2,2} \neq 0$. These surfaces are called the *caustic surfaces* of f.

If $\kappa_{1,1}\kappa_{2,2} = 0$, f is called a *canal surface*. If $\kappa_{1,1} = 0$, then the u^1-parameter curves lie on circles of radius $1/\kappa_1$. In this case, the surface f may be represented as the boundary of the region swept out by a one-parameter family of spheres.

[5] See Strubecker [A15], Vol. III.

3 Surfaces: Local Theory

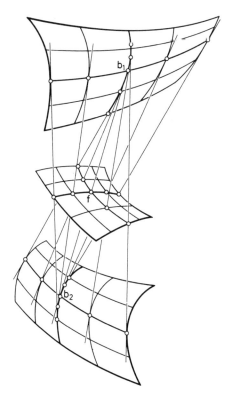

Figure 3.14 Caustic surfaces

3.9.3 Weingarten surfaces.[6] A surface $f\colon U \to \mathbb{R}^3$ is called a *Weingarten surface*, or *W-surface*, provided there exists a function $\varphi\colon U \to \mathbb{R}$ with $d\varphi \neq 0$ such that the principal curvatures $\kappa_1(u) \geq \kappa_2(u)$ satisfy $\varphi(\kappa_1(u), \kappa_2(u)) = 0$. For example, surfaces with $H =$ constant or $K =$ constant are W-surfaces.

Prove: i) On a W-surface, $\kappa_{1,1}\kappa_{2,2} - \kappa_{1,2}\kappa_{2,1} = 0$.

ii) The ellipsoid of revolution,

$$f(u, v) = (a\cdot\cos u\cdot\cos v, a\cdot\cos u\cdot\sin v, b\cdot\sin u)$$

with $0 < a \leq b$, is a W-surface satisfying $\kappa_1 = c\kappa_2$.[5]

iii) A W-surface is not a canal surface if and only if its caustic surfaces consist of asymptotic curves.

3.9.4 A surface with $K = 0$ which is not a developable surface.[7] We will show the existence of a surface $f\colon \mathbb{R} \times\,]-1, 1[\to \mathbb{R}^3$ whose first and second

---

[6] There is a wealth of interesting results about Weingarten surfaces, due to Hilbert, Chern, Hopf, Voss, and others. See, for example, Hopf, H. Über Flächen mit einer Relation zwischen der Hauptkrümmungen. *Math. Nachr.*, **4**, 232–249 (1951). See also Hopf [A9] and [A10].

[7] This example is due to E. Heintze

fundamental forms satisfy

$$(g_{ik}) = (\delta_{ik})$$
$$(h_{ik}) = (P_{ik}(u, v)e^{-((1 \pm v)/u)^2}),$$

where

$$P_{11} = \frac{1}{(1 \pm v)}, \quad P_{12} = \mp \frac{u}{(1 \pm v)^2}, \quad P_{22} = \frac{u^2}{(1 \pm v)^3},$$

and where the sign is the upper when $u \geq 0$ and the lower when $u \leq 0$. We will then show that this surface has zero Gauss curvature but is not a developable surface.
1. The h_{ik} are differentiable.
2. $h_{11}h_{22} - h_{12}^2 = 0$. Therefore $K = 0$.
3. $h_{11,2} = h_{12,1}$, $h_{22,1} = h_{12,2}$.
4. From (2) and (3) one can easily prove that the first and second fundamental forms satisfy the Gauss and Codazzi–Mainardi equations. By the fundamental theorem of surface theory (3.8.8), there exists a surface f with the required first and second fundamental forms. Moreover, f is unique up to an isometry of $\mathbb{R}^3$.
5. The second fundamental form has been chosen so that the inverse image of the generators of f in the set $u < 0$ are the straight lines through $(0, 1)$. In the set $u > 0$, they are the straight lines through $(0, -1)$. The slope of these straight lines blows up as one moves through $(0, 0)$ on the u-axis.
6. The surface f is not a developable surface near $(0, 0)$: there is no change of variables $\phi: V \to U' \subset U$, $(0, 0) \in U'$, such that $f \circ \phi(s, t) = sX(t) + c(t)$.

PROOF. Assume that such a ϕ exists. Without loss of generality, we may assume $\phi(0, 0) = (0, 0)$. Consider the lines parallel to the t-axis in V. They must be mapped into the inverse images of the generators of f which are described in the previous section. Since each of these lines crosses the u-axis exactly once, the inverse image under ϕ of the u-axis may be written in the form $(s, \beta(s))$. If $\rho: \mathbb{R}^2 \to \mathbb{R}^2$ is the map $(\sigma, \tau) \to (\sigma, \beta(\sigma) + \tau)$, the map $\tilde{\phi} = \phi \circ \rho|\rho^{-1}(V)$ is differentiable and $\tilde{\phi}(\sigma, 0) = (a(\sigma), 0)$ for some differentiable function $a(\sigma)$. This follows from the definition of ρ. Therefore $\tilde{\phi}(\sigma, \tau) = (a(\sigma), 0) + \gamma(\tau)(|a(\sigma)|, 1)$, where $\gamma(\tau)$ is a differentiable function with $\gamma(0) = 0$. The function $\tilde{\phi}(\sigma, \tau)$ must have this form because ϕ maps parallels to the τ-axis into the inverse images of the generators.
But $\partial \tilde{\phi}^1/\partial \sigma = a'(\sigma) \pm \gamma(\tau)a'(\sigma)$, the sign depending on the sign of $a(\sigma)$. Since $a(0) = 0$ and $a'(0) \neq 0$, this function cannot be differentiable at any point where $\gamma(\tau) \neq 0$. Contradiction.

3.9.5 Show that the ellipsoid of (3.7.3) with $\rho = u_0 = $ constant $< c$ has exactly four umbilics. In fact the umbilics are precisely the points $x(u_0, v, w)$, $y(u_0, v, w)$, $z(u_0, v, w)$ on the ellipsoid where $v = w = b$. At these points, the lines of curvature are degenerate and grad $\psi(v)$, grad $\psi(w)$ are not defined.

3 Surfaces: Local Theory

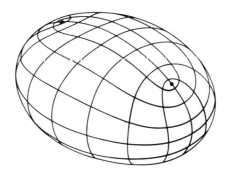

Figure 3.15 Lines of curvature on an ellipsoid. Umbilics marked as dots

3.9.6 A surface $f: U \to \mathbb{R}^3$ is called a *minimal surface* if $H(u) \equiv 0$. The reason for this name is the fact that these are precisely the surfaces for which the first variation of area vanishes. What does this mean?

Consider a family $f^\epsilon(u) = f(u) + \epsilon a(u)n(u)$ of surfaces neighboring f. Here ϵ lies in an interval containing 0 and $a: U \to \mathbb{R}$ is a smooth function. For sufficiently small ϵ, f^ϵ is a regular surface and we may define its first fundamental form. Up to terms of second order and higher in ϵ, $g_{ik}^\epsilon = g_{ik} - 2\epsilon a h_{ik}$, and the *area element* $g^\epsilon = \det(g_{ik}^\epsilon) = g(1 - \epsilon 4aH)$. Therefore $(\partial \sqrt{g^\epsilon}/\partial \epsilon)|_{\epsilon=0} = -2aH$. The only way this can equal zero for all functions a is for H to be identically zero. In (5.6) integration on a surface will be discussed and the area of a surface $f: U \to \mathbb{R}^3$ will be defined as $\int_U \sqrt{g}\, du^1\, du^2$. Using standard techniques of advanced calculus (namely differentiation under the integral sign and the divergence theorem), we may use the above calculation to show that a surface is minimal ($H \equiv 0$) if and only if given any variation $f^\epsilon(u) = f(u) + \epsilon a(u)n(u)$ of f the area function $A(\epsilon) = \int_U \sqrt{g^\epsilon}\, du^1\, du^2 = $ "area of the surface f^ϵ" has a critical point at $\epsilon = 0$. It is easy to see that the area of f cannot be a maximum among nearby surfaces (introducing a pimple on the surface will increase the area). Therefore f must be either a local minimum or some sort of inflection point for the area function.[8]

3.9.7 Consider the surface of revolution f generated by the catenary $(h(u), 0, k(u))$, where

$$h(u) = a \cosh\left(\frac{k(u) - b}{a}\right) \quad \text{(see (3.3.7))}.$$

This surface is known as the *catenoid*.

Prove that the catenoid is the only surface of revolution which is also a minimal surface.

3.9.8 One of the most interesting results in the global theory of minimal surfaces is *Bernstein's theorem*: If $f(u, v) = (u, v, z(u, v))$ is a minimal surface defined for all $(u, v) \in \mathbb{R}^2$, z must be a linear function. In other words, if a minimal surface is the graph of a function defined on the whole plane, then it is a plane.

[8] See Strubecker [A15], Vol. III, p. 222 ff., or the references in footnote 10.

3.9 Exercises and Some Further Results

Figure 3.16 Catenoid

The proof is not elementary, but it is interesting. It utilizes some techniques from complex analysis (see (5.7.4)).[9]

3.9.9 The problem of Plateau.[10] Given a simple, closed, rectifiable curve c in $\mathbb{R}^3$, find a minimal surface $f: D \to \mathbb{R}^3$ spanning c, i.e., if D is the open unit disk, $\bar{D}$ its closure and $S^1 = \partial \bar{D}$ its boundary, does there always exist a continuous function $\bar{f}: \bar{D} \to \mathbb{R}^3$ such that $f := \bar{f} | D$ is a minimal surface and $\bar{f} | S^1: S^1 \to c$ is a homeomorphism, i.e., a continuous, one-to-one mapping onto c with a continuous inverse?

In 1930, T. Rado and J. Douglas independently answered this question in the affirmative. Their solution was not only a minimal surface, but also had minimum possible area among all surfaces $f: D \to \mathbb{R}^3$ which span the given curve c. However, both Rado and Douglas had to admit surfaces with possible isolated singularities. A singularity of a mapping $f: D \to \mathbb{R}^3$ is a point $u \in D$ where df_u has rank < 2. An isolated singularity is a singularity which sits in some neighborhood of all whose points, except u, are not singularities. Whether singularities actually occurred in the Douglas solution to the Plateau problem was an open problem for forty years. In 1970, Osserman was able to show that singularities did not occur in the classical (Douglas) solution to the Plateau problem.[11]

The behavior of $\bar{f}$ at the boundary of $\bar{D}$ was, up until recently, not well understood. Hildebrandt[12] was able to show that if c is differentiable, then

---

[9] There are many proofs of Bernstein's theorem. One of the shortest is due to Nitsche, J. C. C. Elementary proof of Bernstein's theorem on minimal surfaces. *Ann. of Math.*, **66**, 543–544 (1957). For another treatment see Chern [A5].

[10] For a detailed presentation of the solution to the Plateau problem, see Courant, R. *Dirichlet's Problem, Conformal Mappings and Minimal Surfaces.* New York: Interscience Publ., 1950. An excellent introduction to the theory of minimal surfaces in $\mathbb{R}^n$ is Osserman, R. *A Survey of Minimal Surfaces.* New York: Van Nostrand Reinhold, 1969. A compendium of the current knowledge about minimal surfaces can be found in Nitsche, J. C. C. *Vorlesungen Über Minimalfläche,* Springer-Verlag, 1975.

[11] Osserman, R. A proof of regularity everywhere of the classical solution to Plateau's problem. *Ann. of Math.*, **91**, 550–569 (1970).

[12] Hildebrandt, S. Boundary behavior of minimal surfaces. *Arch. Rational Mech. Anal.*, **35**, 47–82 (1969).

$\bar{f}: \bar{D} \to \mathbb{R}^3$, the classical solution to the Plateau problem, is differentiable. The dependence of $\bar{f}$ on c is still an open problem. For example, how many minimal surfaces span a given curve c? What are necessary and sufficient conditions on c which insure the existence of a *unique* solution to the Plateau problem? What conditions on c will insure the existence of an *embedded* solution, i.e., a solution given by a one-to-one mapping f (see Gulliver and Spruck[13]).

One of the ways in which the Plateau problem has been generalized is to seek surfaces of constant mean curvature, $H \equiv \alpha = $ const, spanning a given curve c. Even more generally, one might want the mean curvature H to be specified as a function of position in $\mathbb{R}^3$. One seeks a surface $f: D \to \mathbb{R}^3$ spanning c such that $H(u) = h(f(u))$, where h is a real-valued function defined on $\mathbb{R}^3$. These problems have physical interpretations just as the Plateau problem does. Significant contributions to this subject have been made by Heinz, Hildebrandt, Gulliver, Spruck, and others.[14]

[13] Gulliver, R., and Spruck, J. On embedded minimal surfaces. *Annals of Math.*, **103** (1976), 331–347.

[14] Heinz, E. Über die Existenz einer Fläche konstanter mittlerer Krümmung bei vorgegebener Berandung. *Math. Ann.*, **127**, 258–287 (1954). A useful survey article is Hildebrandt, S. Some recent contributions to Plateau's problem, in *Differentialgeometrie im Grossen*, W. Klingenberg, ed. Mannheim: Bibl. Inst., 1971.

Intrinsic Geometry of Surfaces: Local Theory 4

We are now going to concentrate on the properties of a surface $f: U \to \mathbb{R}^3$ which are *intrinsic* in the sense that they are definable in terms of tangent vectors to the surface and the first fundamental form and its derivatives. For example, the length of a vector or the length of a curve on a surface are intrinsic quantities. The Gauss curvature and the curvature tensor are also intrinsic since they may be defined in terms of the first fundamental form and its derivatives. In contrast, the second fundamental form is not intrinsic. It requires discussion of normal vector fields and cannot, in any case, be reduced to the first fundamental form. Also, principal curvatures are not intrinsic, even though their product, the Gauss curvature, is an intrinsic quantity.

Our point of view will be to use the map $f: U \to \mathbb{R}^3$ to define the first fundamental form as an inner product on $T_u\mathbb{R}^2$, $u \in U$. We have done this previously, but now want to emphasize it. Given $X = \sum_{i=1}^{2} a^i \, \partial f/\partial u^i$, $Y = \sum_{j=1}^{2} b^j \, \partial f/\partial u^j \in T_u\mathbb{R}^2$, $g_u(X, Y) = \sum_{1}^{2} a^i b^j \, \partial f/\partial u^i \cdot \partial f/\partial u^j$. The first fundamental form in this description is an inner product defined on each $T_u\mathbb{R}^2$. As such, we will ultimately want to consider it as given and avoid further reference to the mapping f. In fact, this will be the point of view of the next chapter, in which Riemannian manifolds will be considered without reference to any immersion. For now, we will hold on to the picture of $f: U \to \mathbb{R}^3$ as a surface sitting in Euclidean three-space, using it as a transitional object.

The inner product g_u on $U \subset \mathbb{R}^2$ is not, in general, the standard inner product on $\mathbb{R}^2$. One theme of this chapter will be to generalize familiar properties of the standard inner product on $\mathbb{R}^2$ to new inner products g_u. Of particular interest will be those properties relating to vector differentiation.

4 Intrinsic Geometry of Surfaces: Local Theory

4.1 Vector Fields and Covariant Differentiation

The natural class of vector fields in the study of intrinsic differential geometry of a surface f are the tangential vector fields. These correspond to velocity vectors of paths on $f(U)$. Given a curve $u: I \to U$, it is clear that $f \circ u$ is a curve on $f(U)$. An application of the implicit function theorem, (0.5.2), establishes the converse. Namely, given a regular curve $c(t)$ in $\mathbb{R}^3$ such that $c(t) \subset f(U)$ for all t, then for any t_0 there exists a map $u: I \to U$, defined on a neighborhood of t_0, such that $f \circ u = c$. As a consequence, all tangent vectors to curves on f may be realized as the image under df of tangent vectors to U.

Even if X is a tangential vector field, $\partial X/\partial u^i$ may not be tangential. This partially motivates the next definition.

4.1.1 Definition. Let $f: U \to \mathbb{R}^3$ be a surface, $c = f \circ u: I \to \mathbb{R}^3$ a curve on f, and $X: I \to \mathbb{R}^3$ a tangential vector field along c. For $u \in U$, let $\mathrm{pr}_u: T_{f(u)}\mathbb{R}^3 \to T_u f$ be orthogonal projection in the direction of the normal vector $n(u)$. For $t \in I$, the *covariant derivative* (of X at t), denoted by $\nabla X(t)/dt$, is the vector field $\mathrm{pr}_{u(t)} \circ (dX/dt)(t)$.

4.1.2 The covariant derivative $\nabla X(t)/dt$ is a tangential vector field by definition. Since $dX(t)/dt$ and $\mathrm{pr}_{u(t)}$ are independent of the choice of coordinates, so is $\nabla X(t)/dt$. In terms of a coordinate system (u^1, u^2) on U, we may write $X(t) = \sum_k \xi^k(t) f_{u^k} \circ u(t)$. Then using (3.8.1*),

$$\frac{dX}{dt}(t) = \left(\sum_k \dot{\xi}^k f_{u^k} + \sum_{i,j} \xi^i \dot{u}^j \left(\sum_k \Gamma_{ij}^k f_{u^k} + h_{ij} n \right) \right) \circ u(t).$$

It follows immediately that

$$(*) \qquad \frac{\nabla X}{dt}(t) = \sum_k \left(\dot{\xi}^k + \sum_{ij} \xi^i(t) \dot{u}^j(t) \Gamma_{ij}^k \circ u(t) \right) f_{u^k} \circ u(t).$$

Conclusion: $(\nabla X/dt)(t)$ is an intrinsic geometric quantity whose expression in local coordinates involves the Christoffel symbols.

4.1.3 Lemma. *If $\phi: V \to U$ is a change of variables, let Γ_{ij}^k and $\tilde{\Gamma}_{ij}^k$ be the Christoffel symbols associated with f and $\tilde{f} = f \circ \phi$, respectively. They are related by the following equation:*

$$\tilde{\Gamma}_{pq}^r = \sum_k \frac{\partial^2 u^k}{\partial v^p \partial v^q} \frac{\partial v^r}{\partial u^k} + \sum_{i,j,k} \frac{\partial u^i}{\partial v^p} \frac{\partial u^j}{\partial v^q} \frac{\partial v^r}{\partial u^k} \Gamma_{ij}^k.$$

PROOF. Let ϕ be given in coordinates by $u^i = u^i(v)$. Then

$$X = \sum_r \tilde{\xi}^r \tilde{f}_{v^r} = \sum_k \xi^k f_{u^k}, \quad \text{so} \quad \xi^k = \sum_r \tilde{\xi}^r \frac{\partial u^k}{\partial v^r}, \quad \dot{u}^j = \sum_q \dot{v}^q \frac{\partial u^j}{\partial v^q}.$$

4.1 Vector Fields and Covariant Differentiation

Using (*),

$$\frac{\nabla X(t)}{dt} = \sum_k \left[\sum_r \dot{\xi}^r \frac{\partial u^k}{\partial v^r} + \sum_{p,q} \xi^p \frac{\partial^2 u^k}{\partial v^p \partial v^q} \dot{v}^q \right.$$

$$\left. + \sum_{i,j,r,q} \xi^p \frac{\partial u^i}{\partial v^p} \dot{v}^q \frac{\partial u^j}{\partial v^q} \Gamma^k_{ij} \right] \left[\sum_s \frac{\partial v^s}{\partial u^k} \tilde{f}_{v^s} \right]$$

$$= \sum_r \left[\dot{\xi}^r + \sum_{p,q,k} \xi^p \dot{v}^q \left(\frac{\partial^2 u^k}{\partial v^p \partial v^q} + \sum_{i,j} \frac{\partial u^i}{\partial v^p} \frac{\partial u^j}{\partial v^q} \Gamma^k_{ij} \right) \frac{\partial v^r}{\partial u^k} \right] \tilde{f}_{v^r}$$

$$= \sum_r \left[\dot{\xi}^r + \sum_{p,q} \xi^p \dot{v}^q \tilde{\Gamma}^r_{pq} \right] \tilde{f}_{v^r}.$$

Since this identity must hold for all χ^p and $\dot{v}^q$, the desired result follows. □

4.1.4 Proposition. Let $X(t)$ be two tangential vector fields along $c(t) = f \circ u(t)$. Then

$$\frac{d}{dt} g(X(t), Y(t)) = g\left(\frac{\nabla X(t)}{dt}, Y(t) \right) + g\left(X(t), \frac{\nabla Y(t)}{dt} \right).$$

PROOF. Using the product rule for differentiation,

$$\frac{d(X(t) \cdot Y(t))}{dt} = \frac{dX(t)}{dt} \cdot Y(t) + X(t) \cdot \frac{dY(t)}{dt}.$$

But for $Y \in T_u f$, $Z \cdot Y = (\text{pr } Z) \cdot Y$, where pr is projection onto $T_u f$. The proposition now follows from the definition of covariant differentiation. □

Remark. If $f(u_1, u_2) = (u_1, u_2, 0)$, the surface represented is a piece of the flat plane. Thus $\Gamma^k_{ij} \equiv 0$, $g(X, Y) = X \cdot Y$, and (*) tells us that in this case covariant differentiation is ordinary differentiation.

4.1.5 Let X be a tangential vector field along $f: U \to \mathbb{R}^3$. In coordinates we may write $X(u) = \sum_k \xi^k_{(u)} f_{u^k}(u)$. If $c(t) = f \circ u(t)$ is any curve on f through u_0, $u(0) = u_0$, we may restrict X to $u(t)$ and define $(\nabla X \circ u(t))/dt$, which will have a coordinate representation given by (*) in (4.1.2):

$$\frac{\nabla X \circ u(0)}{dt} = \sum_{j,k} \left(\frac{\partial \xi^k}{\partial u^j}(u_0) + \sum_i \xi^i(u_0) \Gamma^k_{ij}(u_0) \right) \dot{u}^j(0) f_{u^k}(u_0).$$

Notice that the dependence of $(\nabla X \circ u)/dt$ on $u(t)$ involves only the point u_0 and the value of the derivatives $\dot{u}^j(0)$. Consequently, if Y is any tangent vector, $Y \in T_{u_0} f$, $Y = \sum_j \eta_j f_{u^j}(u_0)$, and $c(t) = f \circ u(t)$ is any curve with $u(0) = u_0$, $\dot{c}(0) = Y$, then $(\nabla X \circ u(0))/dt$ will be a vector whose value is independent of the choice of the curve c. We already know by (4.1.2) that $(\nabla X \circ u(0))/dt$ does not depend upon the choice of coordinates on U. Therefore $(\nabla X \circ u(0))/dt$ depends only on the value of $Y \in T_{u_0} f$, and from the form of (*) the dependence is linear.

These observations are summarized below.

4   Intrinsic Geometry of Surfaces: Local Theory

Lemma. *Let X be a tangential vector field on a surface $f: U \to \mathbb{R}^3$. Then for every $u_0 \in U$ we may define a linear map*

$$\nabla X: T_{u_0} f \to T_{u_0} f$$

as follows: If $Y = \sum \eta^j f_j(u_0)$, choose a curve $c(t) = f \circ u(t)$ with $u^j(0) = u_0^j$, $\dot{u}^j(0) = \eta^j$ (for example, let $u^j(t) = u_0^j + t\eta^j$). Then $\nabla X(Y) = (\nabla X \circ u(0))/dt$. The map ∇X is invariantly defined. In particular, $\nabla X(t_0)/dt = \nabla X(\dot{c}(t_0))$.

4.1.6 Definition. Let $X(u)$ be a tangential vector field on f.
  i) ∇X is called the *covariant differential* of X. $\nabla X(Y)$ is the *covariant derivative of X in the direction Y*.
  ii) The function $u \mapsto \operatorname{trace} \nabla X(u)$ from U to $\mathbb{R}$ is called the *divergence of X*, written div $X(u)$.

4.1.7 Observation. Using (4.1.5), we may express div X in coordinates:

$$\operatorname{div} X = \sum_k \frac{\partial \xi^k}{\partial u^k} + \sum_{i,k} \xi^i \Gamma^k_{ik} = \frac{1}{\sqrt{g}} \sum_k \frac{\partial}{\partial u^k}(\sqrt{g}\, \xi^k),$$

where $g = \det(g_{ik})$. In the special case where $f: U \to \mathbb{R}^3$ is a linear and injective map, $f(U)$ is a piece of a plane and $g_{ij}(u) = \delta_{ij}$. Therefore $\Gamma^k_{ij} \equiv 0$ and div $X = \sum_k \partial \xi^k / \partial u^k$. So we see that the divergence of a vector field reduces to the usual notion of divergence when the surface is a piece of a plane. *Note: $(\nabla X \circ u(t))/dt = \nabla X(f \circ u(t))$.*

4.2   Parallel Translation

4.2.1 Covariant differentiation on a surface generalizes ordinary differentiation in the plane. We may now use covariant differentiation to define what it means for vectors or vector fields to be parallel along a curve on a surface. In the plane, a vector field $X(t)$ along a curve $c(t)$ is constant, or parallel, if its value is constant; $X(t) = X_0 = $ constant. In other words, $dX(t)/dt = 0$.

Definition. Let $c = f \circ u$ be a curve on a surface $f: U \to \mathbb{R}^3$. A vector field X along c is *parallel along c* provided $\nabla X(t)/dt = 0$.

4.2.2 It follows immediately from (4.1.4) that if $X(t)$ and $Y(t)$ are both parallel vector fields along c, then $g_{c(t)}(X(t), Y(t))$ is a constant.

Thus a parallel vector field must have constant length and the angle between two parallel vector fields remains constant. Here, in analogy with Euclidean space, the angle between two nonzero vectors X and Y is

$$\theta = \arccos \frac{g(X, Y)}{|X| \cdot |Y|}.$$

4.2.3 Theorem. *Let $f: U \to \mathbb{R}^3$ be a surface, and $c(t) = f \circ u(t)$ a curve on f, $t_0 \leq t \leq t_1$. Let $u(t_0) = u_0$, $u(t_1) = u_1$. Then*

4.2 Parallel Translation

i) *For every $X_0 \in T_{u_0}f$ there exists a unique parallel vector field $X(t)$ along c with $X(t_0) = X_0$.*

ii) *The mapping $\|_c: T_{u_0}f \to T_{u_1}f$, defined by $X_0 \mapsto X(t_1)$, is an isometry.*

PROOF. Suppose $X(t) = \sum \xi^i(t) f_{u^i} \circ u(t)$ is parallel along c. Then $X(t)$ satisfies equation (*) of (4.1.2), namely

$$\dot{\xi}^k(t) + \sum_{i,j} \xi^i(t)\dot{u}^j(t)\Gamma^k_{ij} \circ u(t) = 0, \quad k = 1, 2.$$

But this linear system of two differential equations has a unique solution $\xi^i(t, \xi')$, with initial value $\xi^i(t', \xi') = \xi'^i$, for any $t' \in [t_0, t_1]$. The correspondence $(\xi'^i) \mapsto (\xi^i(t, \xi'))$ is a linear bijection. Finally, we know from (4.2.2) that this map is an isometry. □

Remark. The mapping $\|_c$ generalizes parallel translation in the plane (constant vector fields). Given a vector X_0 at p in the plane, its parallel translation to another point q will be independent of the path c along which we parallel translate. This is not true in general. We will soon see examples of surfaces on which parallel translation is path-dependent.

4.2.4 Technical lemma. *Suppose we are in a coordinate system where $g_{12} = 0$, (orthogonal coordinates). Then $g^{ii} = 1/g_{ii}$, $\Gamma^k_{ik} = (\log \sqrt{g_{kk}})_i = g_{kk,i}/2g_{kk}$, and $\Gamma^k_{ii} = -g_{ii,k}/2g_{kk}$, $(i \neq k)$.*

PROOF. It is easily seen that $g^{11} = 1/g_{11}$, $g^{22} = 1/g_{22}$, and $g^{12} = 0$. Therefore

$$\Gamma^k_{ij} = \tfrac{1}{2} g^{kk}(g_{ik,j} + g_{jk,i} - g_{ij,k}) = \frac{1}{2g_{kk}} \begin{cases} g_{kk,i}, & \text{if } j = k, \\ -g_{ii,k}, & \text{if } j = i, k \neq i. \end{cases} \quad \square$$

4.2.5 An example. *The sphere.* Using the coordinates developed in (3.3.7), $g_{11} = 1$, $g_{12} = 0$, $g_{22} = \cos^2 u$. An application of (4.2.4) yields

$$\Gamma^1_{11} = \Gamma^2_{22} = \Gamma^2_{11} = \Gamma^1_{12} = 0, \quad \Gamma^2_{12} = -\tan u, \quad \Gamma^1_{22} = \cos u \sin u.$$

Consider the curve $c(t) = f(u(t), v(t))$, where $u(t) = a \in \,]-\pi/2, \pi/2[$, $v(t) = t/\cos a$, $0 \le t \le 2\pi \cos a$ (this is the same curve considered in (3.3.7), a latitude circle). The differential equations for the components $\xi^1(t), \xi^2(t)$ of a parallel vector field along c are

$$\dot{\xi}^1(t) + \xi^2(t) \sin a = 0, \quad \dot{\xi}^2(t) - \xi^1(t) \frac{\sin a}{\cos^2 a} = 0.$$

For the initial values $(\xi^1_0, \xi^2_0) = (0, 1/\cos a)$, these equations have the unique solution

$$\xi^1(t) = -\sin(\tan at), \quad \xi^2(t) = \frac{\cos(\tan at)}{\cos a}.$$

In this case we can give an interesting geometric interpretation of parallel translation. In (3.7.7) we showed that the osculating developable to the

77

4 Intrinsic Geometry of Surfaces: Local Theory

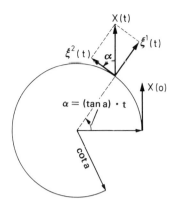

Figure 4.1 Development of the osculating cone

sphere along the latitude circle c was the tangential cone meeting the sphere along c (we will assume that $a \in {]}0, \pi/2[$, so this surface really is a cone). Slitting the cone along the generator through $c(0)$ and applying it to the plane, we consider what happens to the latitude circle $c(t)$ under this transformation. It becomes a circular arc of radius $\cot a$ and length $2\pi \cos a$. At $t = 0$ the tangent vector to this segment is $X(0)$, and $X(t)$ is a parallel (constant) vector field along this arc when considered as a vector field on the plane.

4.2.6 Definitions. Let $c(t) = f \circ u(t)$ be a curve on a surface $f: U \to \mathbb{R}^3$ with $\dot{c}(t) \neq 0$.
 i) The ordered pair of tangential vector fields $e_1(t) := \dot{c}(t)/|\dot{c}(t)|$, $e_2(t)$, along c, where $e_2(t)$ satisfies $|e_2(t)| = 1$, $e_2(t) \cdot e_1(t) = 0$, and $(e_1(t), e_2(t))$ has the same orientation as $(f_{u^1}(u(t)), f_{u^2}(u(t)))$, is called the *Frenet frame of c*.
 ii) $\kappa_g(t) := e_2(t) \cdot ((\nabla e_1(t))/dt)/|\dot{c}(t)|$ is the *geodesic curvature of c*.

Remark. The Frenet frame of a curve c on a surface generalizes the Frenet frame of a plane curve (see (1.4)), and geodesic curvature generalizes the curvature of a plane curve. It is easy to see that Frenet frames are unique. Moreover, the Frenet frame along c and the geodesic curvature of c are invariantly defined with respect to orientation-preserving change of variables. If $t(s)$ is a change of variables and $\tilde{c}(s) = c \circ t(s)$, then $\tilde{\kappa}_g(s) = \pm \kappa_g(t(s))$, the sign being the sign of dt/ds (see (1.3.2)). In the case that $c(t)$ is a unit-speed curve, $|\dot{c}(t)| = 1$, we have $e_1(t) = \dot{c}(t)$ and $\nabla \dot{c}(t)/dt = \kappa_g(t)e_2(t)$. Therefore $\kappa_g(t) = \pm|\nabla \dot{c}(t)/dt|$ (see (1.4)).

4.3 Geodesics

Continuing our study of geometric quantities on surfaces which generalize familiar objects in the plane, we now investigate the analog of straight lines.

4.3.1 Definition. A curve $c(t) = f \circ u(t)$ on a surface $f: U \to \mathbb{R}^3$ is a *geodesic* if $\nabla \dot{c}(t)/dt = 0$.

4.3 Geodesics

4.3.2 Proposition (A characterization of geodesics). *For a regular curve $c(t) = f \circ u(t)$ on a surface f, the following conditions are equivalent:*

i) $\kappa_g(t) = 0$.
ii) *If $s \mapsto t(s)$ is a change of variable on c such that $\tilde{c}(s) = c \circ t(s)$ is a unit-speed curve, then $\tilde{c}(s)$ is a geodesic.*

PROOF

$$\kappa_g(t) = 0 \Leftrightarrow \tilde{\kappa}_g(s) = 0 \quad \text{by (4.2.6)}$$

$$\Leftrightarrow \frac{\nabla \tilde{c}'(s)}{ds} = 0 \quad \text{by (4.2.6).} \qquad \square$$

Remark. Proposition (4.3.2) is the generalization of (1.4.2), which characterizes straight lines in the plane. Notice that it follows immediately from the definition of a geodesic that $|\dot{c}(t)|$ is a constant. Provided that $|\dot{c}(t)| \neq 0$, this means that geodesics are parameterized proportional to arc length. Proposition (4.3.2) says that a regular curve can be reparameterized to be a geodesic if and only if $\kappa_g(t) \equiv 0$. Regular curves satisfying $\kappa_g(t) = 0$ are sometimes called *pre-geodesics*.

In the plane, where $\nabla \dot{c}(t)/dt = d\dot{c}(t)/dt$, it follows that a curve $c(t)$ is a geodesic if and only if $c(t) = At + B$ for some constant vectors A and B. Therefore $c(t)$ is a straight line provided $c(t)$ is regular (and hence $A \neq 0$).

4.3.3 Proposition. *Suppose $c(t) = f \circ u(t)$ is a geodesic. If $u(t) = (u^1(t), u^2(t))$, then $\dot{c}(t) = \sum_k \dot{u}^k(t) f_{u^k} \circ u(t)$, and combining the equations $\nabla \dot{c}(t)/dt = 0$ and (4.1.2) (*), we see that $u(t)$ must satisfy*

$$\ddot{u}^k(t) + \sum_{i,j} \dot{u}^i(t)\dot{u}^j(t) \Gamma^k_{ij} \circ u(t) = 0.$$

Conversely, if $u(t)$ satisfies the above equation, $c = f \circ u(t)$ is a geodesic.

4.3.4 Theorem. *Let $X \in T_{u_0} f$ be a tangent vector to a surface f. Then for sufficiently small $\epsilon > 0$ there exists a unique geodesic $c(t) = f \circ u(t)$, $|t| < \epsilon$, satisfying the initial conditions $u(0) = u_0$, $\dot{c}(0) = X$.*

PROOF. This follows immediately from (4.3.3) and the existence and uniqueness theorem for systems of ordinary differential equations, with initial conditions $u^i(0) = u^i_0$, $\dot{u}^i(0) = \xi^i$, where $X = \sum_i \xi^i f_{u^i}(u_0)$. $\qquad \square$

4.3.5 An example. All the nonconstant geodesics on a sphere ($f = f(u, v)$) of (3.3.7)) are great circles. Recall that

$$f(u, v) = (\cos u \cdot \cos v, \cos u \cdot \sin v, \sin u), \quad (u, v) \in \,]-\pi/2, \pi/2[\, \times \, \mathbb{R}.$$

Since $f_{uu}(u, v) = -f(u, v) = n(u, v)$, $\nabla f_u/du = 0$. Consequently, the $v =$ constant curves, the meridians, are geodesics. Let $c_0(t)$ be one of these meridians with $c_0(0) = f(0, 0)$ and call $\dot{c}_0(0) = X_0$.

Now consider an arbitrary tangent vector $X \in T_{(u_0, v_0)} f$. If $X = 0$, the geodesic with tangent vector X passing through $f(u_0, v_0)$ is the constant

4 Intrinsic Geometry of Surfaces: Local Theory

curve $c(t) = f(u_0, v_0)$. If $X \neq 0$, we might as well assume that $|X| = 1$, since the geodesics through $f(u_0, v_0)$ with initial conditions X or $X/|X|$ are different parameterizations of the same curve.

Now there exists a rotation B of the sphere in $\mathbb{R}^3$ such that $B \circ c_0(0) = f(u_0, v_0)$ and $TBX_0 = X$. Since B leaves the first fundamental form invariant, it must take geodesics into geodesics. Also B takes meridians into great circles on the sphere. Consequently $\tilde{c}(t) = B \circ c(t)$ is a geodesic on the surface $\tilde{f} = B \circ f$ with the initial conditions $c(0) = f(u_0, v_0)$, $\dot{c}(0) = X$. We know $c(t)$ is a great circle.

Of course, $c(t)$ is a curve on the surface $\tilde{f}$ and not on the surface f. In order to conclude that all geodesics on f are great circles, it is now necessary to show that there exists a change of variables ϕ defined on a neighborhood V_0 of (u_0, v_0) with values in a neighborhood U of $(0, 0)$ such that

$$f|V_0 = B \circ f \circ \phi.$$

Then $c(t) = f \circ u(t)$ where $u(t) = \phi^{-1}(0, t)$. We proceed as follows. Since f is regular there exists neighborhoods U_0 of $(0, 0)$ and W_0 of $f(0, 0)$ on the sphere such that $f: U_0 \to W_0$ is a diffeomorphism (see (0.5.2)). Restricting f to a smaller neighborhood if necessary, we may assert that there is a neighborhood V_0 of (u_0, v_0) such that $f|V_0: V_0 \to B(W_0) = B \circ f(U_0)$ is a diffeomorphism. Now let

$$\phi = (f|U_0)^{-1} \circ B^{-1} \circ (f|V_0).$$

It is easy to check that ϕ has the required properties.

Coordinate systems in which some of the coordinate curves are geodesics play an important part in computations as well as in qualitative results in the differential geometry of surfaces.

4.3.6 Lemma (The existence of geodesic orthogonal coordinates). *Let $c(s) = \tilde{f} \circ v(s)$, $s \in I$, be a curve on a surface $\tilde{f}: V \to \mathbb{R}^3$. Fix $s_0 \in \mathring{I}$ and $c'(s_0) \neq 0$. Then there exists a change of variables $\phi: U \to V'$, where V' is an open neighborhood of $v(s_0)$ such that $f = \tilde{f} \circ \phi$ and $u = \phi^{-1} \circ v$ satisfies:*
i) *The curve $c(s) = f \circ u(s)$, for $|s - s_0|$ sufficiently small, is given by $u^1 = 0$, $u^2 = s$.*
ii) *The curves $u^2 = $ constant are geodesics parameterized by arc length. The curves $u^1 = $ constant meet these curves orthogonally. The segment of any $u^2 = $ constant geodesic between the curves $u^1 = a$ and $u^1 = b$ has length $b - a$.*
iii) *The parameters u are an orthogonal coordinate system for f. That is, $g_{12} = 0$. Moreover, $g_{11} = 1$ and, of course, $g_{22} > 0$. Conversely, if the matrix of the first fundamental form satisfies*

$$(g_{ij}) = \begin{pmatrix} 1 & 0 \\ 0 & g_{22} \end{pmatrix},$$

then (ii) is valid.

4.3 Geodesics

In the special case that the initial curve $c(s)$ is a unit-speed geodesic, $g_{22}(0, u^2) = 1$, $g_{22,1}(0, u^2) = 0$, and $\Gamma^k_{ij}(0, u^2) = 0$ for all i, j, k.

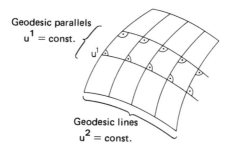

Figure 4.2 Geodesic coordinates

4.3.7 Definition. Coordinates satisfying (ii) or (iii) above are called *geodesic coordinates* (with respect to a curve $u^1 = $ constant). The curves $u^1 = $ constant are called *parallel curves*. If, in addition, the curve $u^1 = 0$ is a geodesic parameterized by arc length then these coordinates are sometimes called *Fermi coordinates*, although they had already been considered by Gauss.

PROOF (of Lemma 4.3.6). 1. Since $c'(s_0) \neq 0$, we may assume, after possibly restricting the domain of definition of c, that $c'(s) \neq 0$ for $s \in \mathring{I}$. This being done, we may assert the existence of the Frenet frame $e_1(s), e_2(s)$. For each $s \in \mathring{I}$ let $c(t, s) = \tilde{f} \circ v(t, s)$ be the geodesic with $c(0, s) = c(s)$ and $(\partial c/\partial t)(0, s) = e_2(s)$. Each of these geodesics is defined for $t < \epsilon(s)$ and by shrinking the domain of definition of $c(s)$ again, if necessary, we may assume that there is an $\epsilon' > 0$ such that $\epsilon(s) > \epsilon'$ for $s \in \mathring{I}$.

2. The mapping $(t, s) \in (-\epsilon', \epsilon') \times \mathring{I} \mapsto (v^1(t, s), v^2(t, s)) \in V$ is differentiable because the $v^i(t, s)$ are solutions to the equation for geodesics and those solutions depend smoothly on the initial conditions $c(s), e_2(s)$, which in turn are differentiable in s. At the point $(0, s_0)$, the matrix of first derivatives of this mapping (the Jacobian matrix) represents vectors which are mapped by df into $e_2(s_0)$ and $c'(s_0)$. Consequently, they are linearly independent. The inverse function theorem, (0.5.1), implies that $\phi(t, s) = (v^1(t, s), v^2(t, s))$ is locally a change of variables.

3. At this point, let us change notation and write (u^1, u^2) instead of (t, s). Now (i) is immediate from the definition of v and u. Also the curves $u^2 = $ constant are unit-speed geodesics by definition. This implies that $g_{11} = 1$ and also that

$$\ddot{u}^k + \sum_{i,j} \dot{u}^i \dot{u}^j \Gamma^k_{ij} = 0 \quad \text{for } u^1 = t, u^2 = \text{constant}.$$

Therefore $\Gamma^1_{11} = \Gamma^2_{11} = 0$. But

$$\Gamma^1_{11} = \frac{1}{2} \sum_l g^{1l}(g_{l1,1} + g_{l1,1} - g_{11,l}) = g^{12} g_{21,1} = 0.$$

Since $g^{12} = -g_{12}/\det(g_{ik})$, the equation above implies $g_{12}g_{21,1} = 0$ or $\frac{1}{2}(g_{12}^2)_{,1} = 0$. Since $g_{12}(0, u^2) = $ (inner product of $e_1(s)$ with $e_2(s)$ along $c(s)) = 0$, it follows that $g_{12} = 0$. Of course, $g_{22} = \det(g_{ik}) > 0$. This proves (ii) and the first part of (iii). (To see that the curves $u^1 = a$ and $u^1 = b$ cut off an arc of length $b - a$ on any geodesic $u^2 = $ constant, simply observe that $u^1 = s$ is arc length on the curve $u^2 = $ constant.)

4. We now prove the second part of (iii). Suppose $g_{11} = 1$, $g_{12} = 0$, and $g_{22} > 0$. By (4.2.4), $\Gamma_{11}^1 = \Gamma_{11}^2 = 0$. Therefore $\nabla f_{u^1}/\partial u^1 = \sum_i \Gamma_{11}^i f_{u^i} = 0$. In other words, the curves $u^2 = $ constant are unit-speed geodesics cutting the curves $u^1 = $ constant orthogonally. Any one of the curves $u^1 = $ constant may serve as basis curve.

5. Suppose $c(s)$ is a unit-speed geodesic. Then

$$0 = \frac{d}{ds}(e_1(s) \cdot e_2(s)) = \frac{\nabla e_1}{ds} \cdot e_2 + \frac{\nabla e_2}{ds} \cdot e_1 = \frac{\nabla e_2}{ds} \cdot e_1.$$

Similarly, $\nabla e_2/ds \cdot e_2 = 0$ since $e_2 \cdot e_2 = 1$. Consequently, $\nabla e_2/ds = 0$. In geodesic coordinates, $e_2(s) = f_{u^1}$ and we may apply (4.1.2) (*) with $(\xi^1, \xi^2) = (1, 0)$. This yields

$$\Gamma_{12}^1(0, u^2) = \Gamma_{12}^2(0, u^2) = 0.$$

By (4.2.4), $\Gamma_{12}^2 = \frac{1}{2}g^{22}g_{22,1}$. Therefore $g_{22,1}(0, u^2) = 0$. Also, $2\Gamma_{22}^2(0, u^2) = g_{22,2}(0, u^2)/g_{22} = 0$ and $2\Gamma_{22}^1(0, u^2) = -g_{22,1}(0, u^2)/g_{11} = 0$. □

4.3.8 Proposition. *In geodesic coordinates, $K(u) = -(\sqrt{g_{22}})_{,11}/\sqrt{g_{22}}$.*

PROOF. By (4.3.6) and (4.2.4), $\Gamma_{11}^1 = \Gamma_{12}^1 = \Gamma_{21}^1 = 0$ and $\Gamma_{12}^2 = (\log \sqrt{g_{22}})_{,1}$. Therefore

$$K = \frac{R_{1212}}{g_{22}} = \frac{g_{22}\Gamma_{11,2}^2 - g_{22}\Gamma_{12,1}^2 + g_{22}(\Gamma_{11}^2\Gamma_{22}^2 - \Gamma_{12}^2\Gamma_{21}^2)}{g_{22}}$$

$$= -(\log \sqrt{g_{22}})_{,11} - ((\log \sqrt{g_{22}})_{,1})^2 = -\frac{(\sqrt{g_{22}})_{,11}}{\sqrt{g_{22}}}. \quad \square$$

If one writes $(\sqrt{g_{22}})_{,11} + K(\sqrt{g_{22}}) = 0$, this turns into a differential equation for $\sqrt{g_{22}}(u^1, u_0^2)$. It will be used below, e.g., in the proofs of (4.4.2) and (4.4.6). Cf. also example (3.9.1).

4.3.9 Theorem. *Let $f: U \to \mathbb{R}^3$ be a surface in geodesic coordinates. Then a geodesic of the form*

$$c = \{c(t) := f(t, u_0^2) \mid t_0 \leq t \leq t_1\}$$

is shorter than any curve $b = \{b(s) := f \circ u(s) \mid s_0 \leq s \leq s_1\}$ from $p_0 = f(t_0, u_0^2)$ to $p_1 = f(t_1, u_0^2)$:

$$L(b) \geq L(c).$$

PROOF

$$L(b) = \int_{s_0}^{s_1} \sqrt{(u^{1'})^2 + g_{22} \circ u(s)(u^{2'})^2}\, ds \geq \int_{s_0}^{s_1} |u^{1'}|\, ds \geq u^1(s_1) - u^1(s_0)$$
$$= t_1 - t_0 = L(c). \quad \square$$

4.4 Surfaces of Constant Curvature

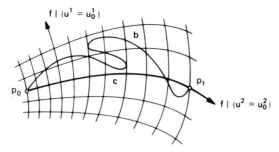

Figure 4.3 A curve in a geodesic coordinate system

Remark. The geodesic $u^2 = u_0^2$ in a geodesic coordinate system is said to be *embedded in a field of geodesics*. In the previous theorem we compared the length of such a geodesic c with a curve b which lies within such a field. If c and b have the same end points, then $L(b) \geq L(c)$. If b does not lie in a field of geodesics, then it is possible that $L(b) < L(c)$. For example, consider a region on the unit sphere of (3.3.7), namely

$$f(u, v) = (\cos u \cdot \cos v, \cos u \cdot \sin v, \sin u), \quad |u| < \pi/4, \ |v| < \pi/2.$$

Using (4.3.5) we see that (u, v) are actually geodesic coordinates based on the curve $v = 0$. However, we may add to this region a patch of surface which meets this piece of a sphere smoothly and joins a neighborhood of $]0, -\pi/2[$ to a neighborhood of $]0, +\pi/2[$ around the back in such a way that it contains a curve b of length approximately 2 which, of course, is strictly less than π. But π is the length of the geodesic $c(t) = f(0, t)$, $-\pi/2 \leq t \leq \pi/2$.

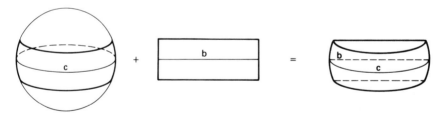

Figure 4.4

4.4 Surfaces of Constant Curvature

4.4.1 Definition. Two surfaces $f: U \to \mathbb{R}^3$ and $\tilde{f}: V \to \mathbb{R}^3$ are *isometric* if there exists a diffeomorphism $\phi: V \to U$ such that

$$g_{\phi(v)}(d\phi X, d\phi Y) = \tilde{g}_v(X, Y)$$

for all $v \in V$ and $X, Y \in T_v \mathbb{R}^2$.

Remark. The map ϕ is called an *isometry*. It is a diffeomorphism which does not stretch the length of vectors or change angles. It is clear that isometry is

4 Intrinsic Geometry of Surfaces: Local Theory

an equivalence relation between surfaces, and that the definition involves only the intrinsic geometry of a surface. If $\phi: V \to U$ is an isometry, then for all $X, Y \in T_v \tilde{f}$,

$$g_{\phi(v)}(df \circ d\phi \circ d\tilde{f}^{-1}X, df \circ d\phi \circ d\tilde{f}^{-1}Y) = \tilde{g}_v(X, Y).$$

To check whether a map ϕ is an isometry, it is only necessary to verify that $g_{\phi(v)}(d\phi \tilde{f}_i, d\phi \tilde{f}_j) = \tilde{g}_v(\tilde{f}_i, \tilde{f}_j) = \tilde{g}_{ij}$. This is because g and $\tilde{g}$ are bilinear and the coordinate vectors form a basis at each point.

An example. The cylinder $f(u, v) = (h(u), k(u), v)$ with $h'^2 + k'^2 = 1$, $(u, v) \in I \times \mathbb{R}$, is isometric to the strip in the plane defined by $\tilde{f}(u, v) = (u, v, 0) \in \mathbb{R}^3$, $(u, v) \in I \times \mathbb{R}$. The map $\phi = \mathrm{id}: I \times \mathbb{R} \to I \times \mathbb{R}$ is an isometry, $g_{ik} = f_i \cdot f_k = \delta_{ik} = \tilde{g}_{ik}$.

Both the cylinder and the plane have zero Gauss curvature. The following theorem will show that this condition characterizes all surfaces which are (locally) isometric to the plane; in other words, all surfaces which may be mapped diffeomorphically onto a piece of the plane without any stretching.

4.4.2 Theorem. *Let $f: U \to \mathbb{R}^3$ be a surface. The following conditions are equivalent.*
   i) $K(u) \equiv 0$.
   ii) *There exist local coordinates in which $g_{ik} = \delta_{ik}$.*
   iii) *Parallel translation is independent of path.*
   iv) *The surface f is locally isometric to an open set of the Euclidean plane $\mathbb{R}^2$.*

Remark. As usual, the use of the word "local" means that the statements hold true for a sufficiently small simply connected neighborhood of any point $u \in U$. In fact, the theorem fails "globally"; the conditions are not equivalent in the large. For example, consider the doubly covered annulus

$$f(u, v) = \left(\frac{u^2 - v^2}{\sqrt{u^2 + v^2}}, \frac{2uv}{\sqrt{u^2 + v^2}}, 0 \right), \qquad 0 < a < u^2 + v^2 < b.$$

Certainly $K = 0$, but (ii) and (iv) fail globally.

PROOF. 0. Notice that (ii) and (iv) are clearly equivalent.

1. (i) $\Rightarrow$ (ii). By using (4.3.6), we may assume that f is presented locally in geodesic coordinates based on a geodesic; so-called Fermi coordinates. The assumption that $K = 0$ implies that $(\sqrt{g_{22}})_{,11} = 0$ (see (4.3.8)). Therefore $(\sqrt{g_{22}})_{,1}$ is a function of the second coordinate only. But since $g_{22,1}(0, u^2) = 0$ in geodesic coordinates, it follows that $g_{22,1} = 0$. Since $g_{22}(0, u^2) = 1$, it must be that $g_{22} = 1$.

In geodesic coordinates, $g_{11} = 1$ and $g_{12} = 0$. Therefore $g_{ij} = \delta_{ik}$.

2. (ii) $\Rightarrow$ (iii). Given (ii) it follows from the coordinate formula for parallel translation, (4.1.2)(*), that parallel translation on f is identical to parallel translation in the plane, and (iii) is true in the plane.

4.4 Surfaces of Constant Curvature

3 (iii) ⇒ (iv). Let (u^1, u^2) be geodesic parallel coordinates based on a geodesic $(0, u^2)$. We wish to show that the curves (u_0^1, u^2) are also geodesics. Consider the unit vector $a = e_2/\sqrt{g_{22}}(u_0^1, 0) \in T_{(u_0^1, 0)}\mathbb{R}^2$. Since the curve $u^2 = 0$ is a geodesic and e_2 is perpendicular to it, the parallel translation of a along $u^2 = 0$ to the point $(0, 0)$ must be a unit vector perpendicular to $u^2 = 0$ at $(0, 0)$. Therefore it is $e_2/\sqrt{g_{22}}(0, 0)$. Since $u^1 = 0$ is a unit-speed geodesic, the parallel translate of $e_2/\sqrt{g_{22}}(0, 0)$ along this curve is simply the tangent vector to this curve. Its value at u_0^2 is $e_2/\sqrt{g_{22}}(0, u_0^2)$. Now parallel translation of this vector along $u^2 = u_0^2$ to (u_0^1, u_0^2) preserves orthogonality and length, so the parallel translate of $e_2/\sqrt{g_{22}}(0, u_0^2)$ at (u_0^1, u_0^2) is $e_2/\sqrt{g_{22}}(u_0^1, u_0^2)$.

Since we are assuming that parallel translation is independent of path, the parallel translate of a along $u^1 = u_0^1$ at the point (u_0^1, u_0^2) must be $e_2/\sqrt{g_{22}}(u_0^1, u_0^2)$. Therefore $e_2/\sqrt{g_{22}}$ is a parallel vector field along $u^1 = u_0^1$. This means that $u^1 = u_0^1$ is a geodesic. Even more, it means that $g_{22}(u_0^1, u^2)$ is a constant function of u^2. Using the geodesic equation of (4.3.3), it follows that $\Gamma^1_{22} = 0$. By (4.2.4), $\Gamma^1_{22} = -g_{22,1}/2g_{11}$. Therefore $g_{22}(u^1, u_0^2) = g_{22}(0, u_0^2) = 1$, since g_{22} is a constant function.

In geodesic parallel coordinates, $g_{11} = 1$ and $g_{12} = 0$, so we now have shown that $g_{ik} = \delta_{ik}$, and (iv) follows from step 0 above.

4. (iv) ⇒ (i). K is invariant under change of variables. So if f is isometric to the plane, then $K = 0$. □

4.4.3 We will now give a geometric interpretation of parallel vector fields along a curve $c = f \circ u$ on a surface f. In (3.7.7), we defined the osculating developable of a surface, and in (3.7.8) and (4.2.5) an example was given which used the osculating developable to interpret parallel translation on the sphere. We will now do this in general. Of course, the osculating developable is not an intrinsic geometric object on a surface, so for the moment we are leaving the realm of intrinsic differential geometry.

Lemma. *Let $c(t) = f \circ u(t)$ be a curve on a surface f. Suppose the osculating developable of f along $c(t)$ is given by*

$$g(s, t) = sY(t) + c(t).$$

If $X(t)$ is a tangential vector field on f along $c(t)$, then $X(t)$ is also a tangential vector field on g along $c(t) = g(0, t)$. Furthermore, $X(t)$ is parallel along c, considered as a curve on f, if and only if $X(t)$ is parallel along c, considered as a curve on g.

4.4.4 Corollary. *The developable surface g is locally isometric to the plane. Therefore $X(t)$ is parallel along $c(t)$ if and only if $X(t)$ is parallel along $c(t)$ in the Euclidean sense when considered as a vector field along a curve in the plane.*

4 Intrinsic Geometry of Surfaces: Local Theory

PROOF. Along $c(t)$ the tangent spaces of f and g agree: $T_{u(t)}f = T_{(0,t)}g$. Therefore $\nabla X/dt = \text{pr } dX/dt$ is the covariant derivative of X along $c(t)$ on both f and g. This proves the lemma. The corollary now follows from (4.4.2) and the fact that g has zero Gauss curvature. □

4.4.5 Examples of surfaces with constant Gauss curvature

1. *The Euclidean plane:* $f(u, v) = (u, v, 0)$ has $K = 0$.
2. *The sphere of radius* $r > 0$: $f^r = (r \cdot \cos u \cdot \cos v, r \cdot \cos u \cdot \sin v, r \cdot \sin u)$ has curvature $K = 1/r^2$. To see this consider geodesic coordinates $\tilde{u} = ur$, $\tilde{v} = v$ based on the equator $u = 0$. Since the equator is a geodesic, these are Fermi coordinates. Let $\tilde{f}(\tilde{u}, \tilde{v}) = f^r(\tilde{u}/r, \tilde{v})$. An easy calculation shows that

$$\tilde{g}_{11} = \tilde{f}_1^2 = 1, \qquad \tilde{g}_{12} = \tilde{f}_1 \cdot \tilde{f}_2 = 0, \qquad \tilde{g}_{22} = \tilde{f}_2^2 = r^2 \cos^2\left(\frac{\tilde{u}}{r}\right),$$

and (4.3.8) allows us to calculate $K = -(\sqrt{g_{22}})_{,11}/\sqrt{g_{22}} = 1/r^2$.
3. *The "pseudosphere"* of (3.9.1) which is the surface of revolution generated by a tractrix:

$$f(u, v) = (h(u) \cos v, h(u) \sin v, k(u))$$

with

$$h(u) = re^{-u/r}, \qquad k(u) = \int_0^u \sqrt{1 - e^{-2t/r}}\, dt, \qquad r > 0,$$

$$f_1^2 = h'^2 + k'^2 = 1, \qquad f_1 \cdot f_2 = 0, \qquad f_2^2 = h^2.$$

These are geodesic parallel coordinates and, by (4.3.8), $K = -1/r^2$.

4.4.6 Proposition. *Suppose f is a surface with Gauss curvature $K = K_0$, a constant. Then in Fermi coordinates*

$$ds^2 = du^2 + \cos^2(\sqrt{K_0}\, u)\, dv^2.$$

Here $\cos(\sqrt{K_0}\, u)$ *is interpreted as* $\cosh(\sqrt{-K_0}\, u)$ *when* $K_0 < 0$.

PROOF. By (4.3.7), $g_{11} = 1$ and $g_{12} = 0$, so in Fermi coordinates based on a geodesic $u = 0$,

$$ds^2 = du^2 + g_{22}\, dv^2$$

with $g_{22}(0, v) = 1$ and $g_{22,1}(0, v) = 0$. We may assume $K_0 \neq 0$, since the case $K_0 = 0$ follows immediately from (4.4.2). By (4.3.8), $(\sqrt{g_{22}})_{,11} + K_0\sqrt{g_{22}} = 0$. With the given initial conditions, this equation has the unique solution

$$\sqrt{g_{22}} = \cos(\sqrt{K_0}\, u).$$
□

We will now use this "normal" form for the line element ds^2 on a surface of constant Gauss curvature to generalize (4.4.2).

4.4.7 Theorem. *Suppose $f: U \to \mathbb{R}^3$ and $\tilde{f}: \tilde{U} \to \mathbb{R}^3$ are two surfaces with constant Gauss curvature. The surfaces f and $\tilde{f}$ have the same constant Gauss curvature if and only if they are locally isometric. Under these conditions, given unit vectors $X_0 \in T_{u_0}f$ and $\tilde{X}_0 \in T_{v_0}\tilde{f}$, there exists a neighborhood U_0 of u_0 and V_0 of v_0 and an isometry $\phi: V_0 \to U_0$ with $\phi(v_0) = u_0$ and $d\phi \circ d\tilde{f}^{-1}\tilde{X}_0 = df^{-1}X_0$.*

PROOF. 1. Suppose f and $\tilde{f}$ have the same constant Gauss curvature. Given $u_0 \in U$ (resp. $v_0 \in V$) and $X_0 \in T_{u_0}f$ a unit vector (resp. $\tilde{X}_0 \in T_{u_0}\tilde{f}$), let $c(t) = f \circ u(t)$ (resp. $\tilde{c}(t) = f \circ v(t)$) be the unit-speed geodesic with $u_0 = u(0)$ and $\dot{c}(0) = X_0$ (resp. $v_0 = v(0)$ and $\dot{c}(0) = \tilde{X}_0$). Introduce Fermi coordinates (u, v) near u_0 based upon the geodesic c (resp. $(\tilde{u}, \tilde{v})$ near v_0 based upon the geodesic $\tilde{c}$). The points $f(u_0)$ and $\tilde{f}(v_0)$ correspond to the coordinate $(0, 0)$. By (4.4.6), the line elements of f and $\tilde{f}$ are in exactly the same form, which means that the local diffeomorphism induced by letting $u = \tilde{u}$ and $v = \tilde{v}$ is a local isometry.

2. Suppose f and $\tilde{f}$ are locally isometric. Then $K(u_0) = K(v_0)$ for every $u_0 \in U$ and $v_0 \in V$. f and $\tilde{f}$ have the same constant Gauss curvature. □

4.5 Examples and Exercises

4.5.1 The geodesics on a surface of revolution.[1] Let f be a surface of revolution as defined in (3.3.7, 3). We will consider those surfaces given in the special form:

$$f(u, v) = (r(u) \cos v, r(u) \sin v, u), \quad r > 0.$$

Recall this is the surface generated by rotating the curve $(r(u), 0, u)$ about the z-axis. The curves $v = v_0 =$ constant are called *meridians*. They are geodesics. The curves $u = u_0 =$ constant are called *parallel circles*. They are circles of radius equal to $r(u_0)$.

Let $T^0 f$ denote the collection of nonzero tangent vectors on f. If $X \in T^0 f$, define $\theta(X)$ to be the angle between X and the parallel circle $u = u_0$ (here $X \in T_{(u_0, v_0)}f$), i.e.,

$$\theta(X) := \arccos(X \cdot f_v(u_0, v_0)/|X|r(u_0)).$$

The mapping

$$\Phi: T^0 f \to \mathbb{R},$$

defined by $X \mapsto r(u_0) \cos \theta(X)$, determines almost all the geodesics on f. Prove the following theorem due to Clairaut: A curve $c(t) = f(u(t), v(t))$ on f which satisfies $\dot{u}(t) \neq 0$ is a pre-geodesic if and only if $\Phi(\dot{c}(t))$ is a constant. (This theorem, which expresses the conservation of the angular momentum Φ, is a special case of a more general result about surfaces which may be expressed in local coordinates whose line element has a specific form (Liouville line element). See (5.7.5).)

[1] See Darboux [A6], Volume III, Book 6, Chapter 1.

4   Intrinsic Geometry of Surfaces: Local Theory

Clairaut's theorem enables us to give a qualitative description of the geodesics on a surface of revolution. To simplify matters, let us assume that the surface f possesses an "equator." By this we mean that $r(u) \leq r(0)$ with equality if and only if $u = 0$, and for every $u_+ > 0$ in the domain of definition there exists a unique $u_- < 0$ such that $r(u_+) = r(u_-)$. In other words, to every northern latitude circle there corresponds exactly one southern latitude circle and conversely. This boils down to an assumption about the shape of the meridian curve; in particular, $r(u)$ must have a strict local maximum at $u = 0$.

Let θ_0 be an angle small enough to insure the existence of a pair u_+, u_- in the u-parameter interval such that $r(0) \cos \theta_0 = r(u_+) = r(u_-)$.

Show: (i) There exists a geodesic which (a) cuts the equator at an angle of θ_0, (b) crosses every parallel circle $u = $ constant for $u_- \leq u \leq u_+$, (c) lies entirely in the region of the surface of revolution with $u_- \leq u \leq u_+$, and (d) meets the parallel circles $u = u_+$ and $u = u_-$ tangentially.

Since rotation, $u \mapsto u$, $v \mapsto v + v_0$, is an isometry of a surface of revolution, the above result characterizes every geodesic which crosses the equator at a sufficiently shallow angle.

(ii) The equator itself is a geodesic. More generally, on any surface of revolution a parallel circle $u = u_0 = $ constant is a geodesic if and only if $r'(u_0) = 0$.

4.5.2 Examples of surfaces of revolution with an equator.[1] The surfaces of revolution with constant curvature $K = 1$ of (3.9.1, ii) all have equators of length $2\pi a$. By using the fact that these surfaces are locally isometric to the sphere of constant curvature $K = 1$ (for which $a = 1$), show: (i) If a is irrational, a geodesic which crosses the equator making a sufficiently small angle θ_0 (small enough so that the geodesic is defined for all values of t, see (4.5.1)) will never close up smoothly. Consequently, the equator is an isolated closed geodesic. (ii) If a is rational, i.e., $a = p/q$ with p and q relatively prime, then all geodesics which cross the equator making a sufficiently small angle $\theta_0 \neq 0$ must be smoothly closed curves of length $2\pi q$. Consequently, any small perturbation of the initial conditions defining the equatorial geodesic will be the initial conditions of a closed geodesic.

[1] See Darboux [A6], Volume III, Book 6, Chapter 1.

Two-Dimensional Riemannian Geometry 5

In the previous chapter, we considered the intrinsic geometry of a surface $f: U \to \mathbb{R}^3$. Many *geometric* properties of surfaces were presented in terms of the open set U, together with the positive definite inner product g_u on each $T_u\mathbb{R}^2$ (i.e., in terms of the first fundamental form). The *geometric* properties were those invariant under change of variable.

We did, however, continue to distinguish between surfaces which were isometric but not congruent. For example, we made a distinction between the cylinder and the plane in (4.4). The cylinder is locally isometric to the plane, but there does not exist an isometry of $\mathbb{R}^3$ which maps the plane into the cylinder, even locally. This distinction is not an intrinsic one, and involves reference to the ambient space $\mathbb{R}^3$, and to the respective mappings which define the plane and the cylinder.

In this chapter, we will make two important generalizations of the notion of a surface. First, a (local) surface will be defined to be an open set $U \subset \mathbb{R}^2$, together with a positive definite inner product g_u on each $T_u\mathbb{R}^2$. The inner product is not required to be derived from some $f: U \to \mathbb{R}^3$. It is only required to be differentiable as a function of $u \in U$. Second, the idea of a manifold will be introduced. A two-dimensional manifold is a topological space which, locally, is homeomorphic to an open set in $\mathbb{R}^2$. For example, each point on the sphere S^2 in $\mathbb{R}^3$ has a neighborhood homeomorphic to an open set in $\mathbb{R}^2$, but the entire manifold S^2 does not have this property. We will want to consider manifolds on which a positive definite inner product is defined at each point, i.e., Riemannian manifolds.

5 Two-Dimensional Riemannian Geometry

5.1 Local Riemannian Geometry

Let $S(2)$ denote the set of all real symmetric, positive definite 2×2 matrices (g_{ik}). An element of $S(2)$ corresponds to the matrix representation of a positive definite quadratic form on the vector space $\mathbb{R}^2$ (see (3.2.1)). As a set, $S(2)$ may be considered an open subset of the three-dimensional space of all 2×2 symmetric matrices, and as such we may speak of *differentiable* maps from $\mathbb{R}^2$ into $S(2)$, meaning that the induced map from $\mathbb{R}^2$ to $\mathbb{R}^3$ is differentiable.

5.1.1 Definitions. i) Let U be an open subset of $\mathbb{R}^2$. A *Riemannian metric on U* is a differentiable map

$$g: U \to S(2).$$

Notation: We will denote a Riemannian metric on U by (U, g).

If (U, g) and $(V, \tilde{g})$ are two sets with Riemannian metrics, they are *equivalent* if they are isometric. In other words, they are equivalent if there exists a diffeomorphism $\phi: V \to U$ such that

$$g_{\phi(v)}(d\phi X, d\phi Y) = \tilde{g}_v(X, Y) \quad \text{for all } X, Y \in T_v\mathbb{R}^2 \text{ and all } v \in V.$$

If (U, g) and $(V, \tilde{g})$ are equivalent via an orientation-preserving diffeomorphism ϕ (det $d\phi > 0$), they are said to be *positively equivalent*.

ii) A *(local) surface with Riemannian metric* is an equivalence class of sets with Riemannian metric.

A *(local) oriented surface with Riemannian metric* is a class of sets with a Riemannian metric which are positively equivalent.

We will use M to denote one of these equivalence classes. In general, M will be written in terms of one of the (U, g) and we will call (U, g) a *coordinate system of M*. The elements of U will correspond to *points* of M and these points will be denoted by the letters $p, q, r, \ldots$.

Remark. If $f: U \to \mathbb{R}^3$ is a surface in the sense of Chapters 3 and 4, it defines a surface with a Riemannian metric, namely the equivalence class of (U, g) with $g_u = I_u$.

We will now prove that all the geometric objects of Chapter 4, which may be defined in terms of the Riemannian metric $g_u = I_u$ and which are invariant under change of variables, may be generalized to geometric objects on a surface with Riemannian metric. To wit:

5.1.2 Lemma. *Let M be a surface with a Riemannian metric. Let (U, g) be a coordinate system for M.*

5.1 Local Riemannian Geometry

i) Let c be a curve on M, represented by $u(t)$, $t_0 \leq t \leq t_1$. Then the length of c, $L(c)$, and the energy of c, $E(c)$, defined by

$$L(c) := \int_{t_0}^{t_1} \sqrt{\sum_{i,k} g_{ik} \circ u(t) \dot{u}^i(t) \dot{u}^k(t)} \, dt$$

$$E(c) := \frac{1}{2} \int_{t_0}^{t_1} \sum_{i,k} g_{ik} \circ u(t) \dot{u}^i(t) \dot{u}^k(t) \, dt,$$

are invariantly defined.

ii) Define the Christoffel symbols Γ_{ij}^k by

$$\Gamma_{ij}^k := \frac{1}{2} \sum_{l} g^{kl}(g_{li,j} + g_{lj,i} - g_{ij,l})$$

and the covariant derivative of the basis vector fields $e_i(u)$ by

$$\frac{\nabla e_i(u)}{\partial u^j} := \sum_k \Gamma_{ij}^k \circ u \, e_k(u), \quad 1 \leq i, j, k \leq 2.$$

If X is a vector field on M, the covariant differential ∇X, and the divergence div X, may be defined as follows. In terms of the coordinates (U, g), X may be written as $\sum_k \xi^k(u) e_k(u)$. Then $\nabla X \colon T_u \mathbb{R}^2 \to T_u \mathbb{R}^2$ is the linear transformation corresponding to the matrix

$$(\nabla X(u)_j^k) = \left(\frac{\partial \xi^k(u)}{\partial u^j} + \sum_i \xi^i(u) \Gamma_{ij}^k \circ u \right)$$

(see (4.1.5)) and

$$\operatorname{div} X = \operatorname{trace} \nabla X = \frac{1}{\sqrt{g}} \sum_k \frac{\partial}{\partial u^k} (\sqrt{g} \xi^k),$$

where $g = \det(g_{ik})$. (Compare with (4.1.7, 1)).

All of these quantities are invariantly defined.

iii) The covariant derivative $\nabla X(t)/dt$ of a vector field $X(t)$ along a curve $c(t)$ in M may be defined in terms of a coordinate system (U, g) by using the formula (4.1.2(*)). Let $X(t) = \sum_k \xi^k(t) e_k \circ u(t)$. Then

$$\frac{\nabla X(t)}{dt} = \sum_k \left(\dot{\xi}^k(t) + \sum_{i,j} \xi^i(t) \dot{u}^j(t) \Gamma_{ij}^k \circ u(t) \right) e_k \circ u(t).$$

Using this definition, we may now speak of parallel vector fields $X(t)$ along $c(t)$, i.e., vector fields satisfying $\nabla X(t)/dt = 0$.

iv) The Frenet frame of a regular curve $c(t)$ on an orientable surface M is definable exactly as in (4.2.6).

v) Geodesics as in (4.3.1).

vi) The curvature tensor, defined on (U, g) as in (3.8.4), is coordinate invariant. It is given by

$$R_{iljk} = g\left(\frac{\nabla}{\partial u^k} \frac{\nabla}{\partial u^j} e_i - \frac{\nabla}{\partial u^j} \frac{\nabla}{\partial u^k} e_i, e_l \right).$$

vii) *The Gauss curvature K is invariantly defined. With respect to (U, g), it is $K = R_{1212}/\det(g_{ij})$.*

PROOF. The above definitions involve tangent vectors, curves, and the Riemannian metric, all of which may be expressed in terms of a local coordinate system, (U, g). What needs to be verified is that these definitions are independent of choice of coordinate system.

1. Suppose $\phi: (V, \tilde{g}) \to (U, g)$ is an isometry. This means that

$$\tilde{g}_{ij}(v) = \sum_{k,l} \frac{\partial u^k}{\partial v^i} \frac{\partial u^l}{\partial v^j} g_{kl}(\phi(v)).$$

From this it is clear that length and energy are invariant under change of coordinates.

2. To show that the expression (4.1.2(*)) for the covariant derivative is coordinate invariant, it suffices to verify the transformation law (4.1.3) for the Christoffel symbols. This may be done by direct calculation. If such a calculation is not to your taste, here is an alternate proof. First express $\tilde{g}_{pq}$ and $\tilde{\Gamma}^r_{pq}$ in terms of g_{ik} and Γ^l_{ik}. Now consider (4.1.3) as an identity in which the Γ^k_{ij} appear linearly, with coefficients of the form $\partial u^i/\partial v^p$, $\partial^2 u^i/\partial v^p \partial v^q$ and their products.

We now claim that, given $u_0 \in U$, there is a surface $f: U \to \mathbb{R}^3$ such that the $g_{ij}(u_0)$ and $\Gamma^k_{ij}(u_0)$, defined by f, agree with those given by the Riemannian metric on U at u_0. We have already verified (4.1.3) for surfaces $f: U \to \mathbb{R}^3$, and the identity will then follow in the Riemannian case.

To prove the claim, observe first that it is certainly possible to construct an f with the required $g_{ij}(u_0)$. We may then introduce a change of variables $\phi: (v^1, v^2) \mapsto (u^1, u^2)$ with $\partial u^k/\partial v^i = \delta^k_i$ at u_0 and $(\partial^2 u^k/\partial u^p \partial u^r)(u_0)$ arbitrary. Using the transformation law for the Christoffel symbols, (4.1.3), it follows that for an appropriate choice of ϕ the mapping $\tilde{f} = f \circ \phi$ will have the required Christoffel symbols.

From this (ii) and therefore (iii)–(vii) follow: all the quantities are parameter-invariant. The only loose end is the invariance of R_{iljk}, but this follows directly from the definition of R_{iljk}. □

Before continuing with our general development of the subject of surfaces with a Riemannian metric, let us pause to consider a very important example.

5.1.3 The hyperbolic plane (the Poincaré half-plane) H^2_r.

The surface H^2_r is the set $U := \{(u, v) \in \mathbb{R}^2 \mid v > 0\}$, together with the Riemannian metric $ds^2 := (r^2 du^2 + r^2 dv^2)/v^2$, $r > 0$ (see (3.4.2)). Recall that this notation for the metric is equivalent to $g_{ik} = (r^2/v^2)\delta_{ik}$.

Introduce geodesic coordinates based on a horizontal line $v = v_0 > 0$ as follows: $u = \tilde{v}$, $v = \exp(-\tilde{u}/r)$, $(\tilde{u}, \tilde{v}) \in \mathbb{R} \times \mathbb{R}$. Computing $(\tilde{g}_{ik})$, using the transformation law for the first fundamental form under change of variables, we get $\tilde{g}_{11} = 1$, $\tilde{g}_{12} = 0$, $\tilde{g}_{22} = r^2 \exp(2\tilde{u}/r)$. Therefore, by (4.3.8), $K =$

$(-\sqrt{\tilde{g}_{22}})_{,11}/\sqrt{\tilde{g}_{22}} = -1/r^2$. The hyperbolic plane H_r^2 has constant Gauss curvature equal to $-1/r^2$.

Notice that the line element ds^2 of H_r^2 in the (u, v) coordinates is equal to the Euclidean line element $du^2 + dv^2$ multiplied by a function, i.e., it is proportional to the Euclidean line element. Because of this, angles measured in the Euclidean upper half-plane are equal to angles measured in the metric of H_r^2.

Remark. Given two surfaces with Riemannian metric (U, g) and $(\tilde{U}, \tilde{g})$, a mapping $\phi: U \to \tilde{U}$ is *conformal* if $\tilde{g}_{\phi(u)}(d\phi(X), d\phi(Y)) = \lambda(u)g_u(X, Y)$ for all $X, Y \in TU$. Here $\lambda: U \to \mathbb{R}$ is a real strictly positive differentiable function. It is straightforward to prove that if ϕ is conformal, ϕ preserves angles. In the above example, the identity is a conformal mapping.

The hyperbolic plane provides a negative answer to a very natural question that may have already occurred to the astute reader. Is it true that every surface with Riemannian metric (U, g) can be realized as a surface $f: U \to \mathbb{R}^3$? (That is, the metric induced by the mapping f is isometric to g.) In 1901, David Hilbert proved that H_r^2 cannot be realized as a surface in $\mathbb{R}^3$.[1] Nonetheless, each point $p \in H_r^2$ has a neighborhood V which may be realized as a surface $f: V \to \mathbb{R}^3$. In fact, we have all but proved this already. The pseudosphere of (4.4.5) is a surface in $\mathbb{R}^3$ with constant negative Gauss curvature $-1/r^2$. But Theorem 4.4.7 says any two such surfaces of the same constant Gauss curvature are locally isometric.

There is not a globally defined isometry, however. Briefly, H_r^2 is *simply connected* and *complete* (for precise definitions, see (6.6.2) and (6.4.4), respectively) and the pseudosphere is neither. Any global isometry would preserve these properties. A proof of Hilbert's nonexistence theorem may be found in Hopf [A11] or do Carmo [A8].

5.1.4 A brief word about transformation groups. Let E be a set and G a group. The group G *acts on E as a transformation group* if there exists a mapping $G \times E \to E$; $(g, x) \mapsto gx$ such that

$$(g_1 g_2)x = g_1(g_2 x)$$

and

$$ex = x, \quad \text{where } e \in G \text{ is the identity element.}$$

For each $g \in G$, the map $g: E \to E$; $x \mapsto gx$ is a bijection since g^{-1} is its inverse. Of course, a group G may act on a set E in more than one way.

An action of G on E is *transitive* provided that for each pair $x_1, x_2 \in G$ there exists a $g \in G$ such that $gx_1 = x_2$.

Given $x \in E$, the *isotropy subgroup* G_x is the set of all $g \in G$ such that $gx = x$. It is easy to check that G_x is in fact a subgroup.

---
[1] Hilbert, D. Über Flächen von konstanter Gausscher Krümmung. *Trans. Amer. Math. Soc.* **2**, 87–99 (1901). For further references, see Nirenberg [A12].

5 Two-Dimensional Riemannian Geometry

5.1.5 Definition. *SL(2, ℝ)*, the *special linear group in dimension* 2, is the group of all real (2 × 2)-matrices with determinant = 1.

We may define an action of *SL(2, ℝ)* on H_r^2 as follows. First introduce the complex variable $z = u + iv$. The points (u, v) in the upper half-plane correspond to $z = u + iv$, $v > 0$. Given $g = \begin{pmatrix} a & b \\ c & d \end{pmatrix} \in SL(2, \mathbb{R})$, let $gz = (az + b)/(cz + d)$.

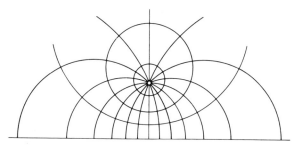

Figure 5.1 Geodesic circles in the Poincaré half-plane

It is easy to verify that $(g, z) \mapsto gz$ is an action of *SL(2, ℝ)* on H_r^2. In fact:

5.1.6 Proposition. *The group SL(2, ℝ) acts as a group of isometries on H_r^2. Moreover, the action is transitive (even stronger, given any two unit tangent vectors to H_r^2, there exists a $g \in SL(2, \mathbb{R})$ such that dg maps one into the other). The isotropy subgroup of any point of H_r^2 is isomorphic to SO(2), the group of rotations of the Euclidean plane.*

PROOF. 1. Let $u + iv = z$ and $(az + b)/(cz + d) = \tilde{z}$. If we write $dz\,d\bar{z}$ for $du^2 + dv^2$, the line element for H_r^2 at z may be written $ds^2(z) = -4r^2\,dz\,d\bar{z}/(z - \bar{z})^2$. (Recall $\bar{z} = u - iv$). An easy calculation shows that $d\tilde{z} = d((az + b)/(cz + d)) = dz/(cz + d)^2$ and therefore $ds^2(z) = ds^2(\tilde{z})$. This means that $z \mapsto \tilde{z}$ is an isometry.

2. If $z = i$, then $\tilde{z} = \tilde{u} + i\tilde{v} = (ai + b)/(ci + d) = (bd + ac)/(c^2 + d^2) + i(1/(c^2 + d^2))$. Now, given any $(\tilde{u}, \tilde{v})$ with $\tilde{v} > 0$, there exists a $g = \begin{pmatrix} a & b \\ c & d \end{pmatrix}$, with $ad - bc = 1$, such that g maps $(0, 1)$ into $(\tilde{u}, \tilde{v})$. Namely, let $d = 0$, $c = 1/\sqrt{\tilde{v}}$, $a = \tilde{u}/\sqrt{\tilde{v}}$, and $b = -\sqrt{\tilde{v}}$. Therefore *SL(2, ℝ)* acts transitively on H_r^2.

3. The isotropy group of $z = i$ is the group of all matrices $\begin{pmatrix} a & b \\ c & d \end{pmatrix}$ with $bd + ac = 0$, $c^2 + d^2 = 1$, and $ad - bc = 1$. This implies that, for some $\phi \in [0, 2\pi]$, $a = d = \cos\phi$ and $b = -c = \sin\phi$. Conversely, given $\phi \in [0, 2\pi]$,

$$\begin{pmatrix} \cos\phi & -\sin\phi \\ \sin\phi & \cos\phi \end{pmatrix}$$

is an element of the isotropy group of $z = i$. Therefore $SL(2, \mathbb{R})_i = SO(2)_i$, i.e., the isotropy group of *SL(2, ℝ)* at i is *SO(2)*.

The isotropy groups of any z and z' are conjugate to one another. For, if $g \in SL(2, \mathbb{R})$ takes z to $gz = z'$, then $G_{z'} = gG_z g^{-1}$. Therefore all isotropy groups of this action are isomorphic to *SO(2)*. Combining this

94

result with (2) above also proves that $SL(2, \mathbb{R})$ acts transitively on the unit tangent vectors of H_r^2. □

5.1.7 The geodesics on H_r^2 are, modulo parameterization, circles or straight lines (in the Euclidean sense) which meet the boundary $v = 0$ orthogonally. To prove this it is sufficient to establish the result in the case $r = 1$, since the identity map from H_1^2 to H_r^2 is a homothetic transformation with constant $= r$ (i.e., a conformal map with $\lambda(u) \equiv r$). Such a map must preserve geodesics. (Proof: exercise.)

In H_1^2, $g_{11} = 1/v^2$, $g_{12} = 0$, and $g_{22} = 1/v^2$. Therefore $\Gamma_{11}^1 = \Gamma_{22}^1 = \Gamma_{12}^2 = 0$ and $\Gamma_{11}^2 = -\Gamma_{22}^2 = -\Gamma_{21}^1 = 1/v$. The differential equations for geodesics, (4.3.3), can therefore be written in the form

$$\ddot{u} - \frac{2\dot{u}\dot{v}}{v} = 0, \qquad \ddot{v} + \frac{\dot{u}^2 - \dot{v}^2}{v} = 0.$$

If $\dot{u} = 0$, then $u = $ constant. In this case the geodesic is a line orthogonal to $v = 0$.

If $\dot{u} \neq 0$, the first equation implies that $\ln(\dot{u}/v^2) = $ constant and therefore $\dot{u} = cv^2 \neq 0$ for some constant c. Similarly, the second equation implies $\dot{u}^2 + \dot{v}^2 = bv^2 > 0$ for some constant b. Combining these two equations gives $(dv/du)^2 = \dot{v}^2/\dot{u}^2 = b/c^2 v^2 - 1$. Therefore $(u - a)^2 + v^2 = b/c^2$ for some constant a. This is a circle with center on $v = 0$. Hence the circle meets $v = 0$ orthogonally.

5.2 The Tangent Bundle and the Exponential Map

The notion of the tangent bundle TU of $U \subset \mathbb{R}^2$ was introduced in (0.4). We recall briefly some notation and basic facts. First, $\pi = \pi_U : TU \to U$ denotes the projection. The inverse image $\pi^{-1}(u)$ of u is precisely $T_u\mathbb{R}^2$. The canonical identification $TU \cong U \times \mathbb{R}^2$ allows us to define a differentiable structure on TU (i.e., as a subset of $\mathbb{R}^4$) and therefore it makes sense to speak of differentiable functions $f: TU \to \mathbb{R}$ or differentiable mappings $X: U \to TU$.

Suppose now that (U, g) and $(V, \tilde{g})$ are two coordinate systems for a surface M. There must be an isometry $\phi: V \to U$, that is, a diffeomorphism with $g(d\phi, d\phi) = \tilde{g}(\_, \_)$. The tangential of ϕ,

$$T\phi: TV \to TU,$$

must also be a diffeomorphism (for definition, see (0.4)). Moreover, $T\phi$ is compatible with the projections; $\pi_U \circ T\phi = \phi \circ \pi_V$. Also, $T\phi|T_v\mathbb{R}^2$ maps $T_v\mathbb{R}^2$ onto $T_{\phi(v)}\mathbb{R}^2$ isometrically. Using this we may make the following definitions.

5.2.1 Definitions. Let M be a surface with a Riemannian metric.
i) Let $(V, \tilde{g})$ and (U, g) be representations of M and $\phi: V \to U$ an isometry. We will say that $X_v \in T_v\mathbb{R}^2 \subset TV$ is equivalent to $X_u \in T_u\mathbb{R}^2 \subset TU$ provided $T\phi(v, X_v) = (u, X_u)$, i.e., $\phi(v) = u$ and $d\phi_v X_v = X_u$. A *tangent vector* to M is an equivalence class of such vectors.

5 Two-Dimensional Riemannian Geometry

ii) Every tangent vector X to M determines an element of M. If X is represented by $(u, X_u) \in T_u\mathbb{R}^2 \subset TU$, the point $p \in M$ represented by $u \in U$ is called the *base point of the tangent vector* X. The base point of X is defined independently of choice of coordinate systems.

iii) The *tangent bundle* of M, denoted by TM, is the set of all tangent vectors of M, together with the map $\pi: TM \to M$ which maps $X \in TM$ to its base point. If U is a representative of M, TU together with $\pi_U: TU \to U$ is called a representation of TM. The tangent bundle TM of M has a natural differentiable structure inherited from the differentiable structure of its representatives. This differentiable structure is clearly independent of the choice of representative.

iv) The inverse image $\pi^{-1}(p)$ of a point $p \in M$ under the bundle projection $\pi: TM \to M$ is called the *tangent space of M at p*. Notation: T_pM. The space T_pM consists of precisely those vectors in TM with base point equal to p. If TM is represented by TU and p is represented by u, T_pM is represented by $T_u\mathbb{R}^2$. Via this identification, T_pM has the structure of a two-dimensional real vector space with a positive definite inner product g_p defined by g_u.

v) Given $X \in TM$, the norm of X, $|X|$, is defined by $|X| := |X_u| := \sqrt{g_u(X_u, X_u)}$, where $X_u \in T_u\mathbb{R}^2$ is a representative of X.

vi) Let $\epsilon > 0$. By $B_\epsilon M$ we will mean the set of all $X \in TM$ with $|X| < \epsilon$. The set $B_\epsilon \subset TM$ is an open set because, in a representation TU, $B_\epsilon M$ is represented by the set of all X_u with $|X_u| < \epsilon$. This set is the inverse image of the open interval $]-\epsilon, \epsilon[$ under the continuous function $v: TU \to \mathbb{R}$ that carries X_u into $|X_u|$.

Remark. For each $p \in M$, $B_\epsilon M \cap T_pM$ is the open disc $B_\epsilon(0)$ centered at the origin in T_pM. Let $X \in T_pM$. Given a sufficiently small $\epsilon > 0$ (ϵ depends on X), there exists a unique geodesic $c(t)$, $|t| < \epsilon$, in M with $\dot{c}(0) = X$ (this follows from (4.3.4)). We shall denote this geodesic by c_X.

We now want to use this fact in order to construct a map from $B_\epsilon(0)$ onto a neighborhood of p in M. To be precise, the map we will use is $X \in B_\epsilon(0) \to c_X(1)$. Even more, we would like to do this simultaneously for all $p \in M$ in a sufficiently small neighborhood of a point $p_0 \in M$.

5.2.2 Lemma. *Let M be a surface with a Riemannian metric and let $p_0 \in M$. Then there exists an open neighborhood M_0 of p_0 and an $\epsilon = \epsilon(p_0) > 0$ such that the map $B_\epsilon M_0 \to M$ given by $X \to c_X(1)$ is defined and differentiable. Consequently, if $\pi X \in M_0$ and $|X| < \epsilon$, then tX, $0 \leq t \leq 1$, gets mapped into a geodesic $c_X(t)$.*

PROOF. 1. We will do everything in a coordinate system (U, g) of M. The point p_0 will be represented by u_0.

5.2 The Tangent Bundle and the Exponential Map

2. The differential equations (4.3.3)(*) for a geodesic can be written in the form

$$\dot{u}^k = v^k, \qquad \dot{v}^k = -\sum_{i,j} v^i v^j \Gamma_{ij}^k(u).$$

Let $u(t; u, X)$, $v(t; u, X)$ be the solution of these equations which have the initial value $(u, X) \in U \times \mathbb{R}^2 \cong TU$ when $t = 0$. Applying well-known theorems of the theory of ordinary differential equations (see Hurewicz, Lectures on Ordinary Differential Equations, M.I.T. Press, 1958), there exists a neighborhood $W = \,]-2\theta, 2\theta[\, \times B_{2\delta}^e(u_0) \times B_\eta^e(0)$ of $(0, u_0, 0) \in \mathbb{R} \times U \times \mathbb{R}^2$ on which the map $\Phi: W \to U \times \mathbb{R}^2$ given by

$$(t, u, X) \mapsto (u(t; u, X), v(t; u, X))$$

is differentiable. Here B_ρ^e denotes the disk of radius ρ in the Euclidean metric.

3. Since $B_\delta^e(u_0)$ is relatively compact in U, there exists a $\gamma > 0$ such that, for every $u \in B_\delta^e(u_0)$, $X \cdot X = (x_1^2 + x_2^2) \leq \gamma^2 g_u(X, X)$.

   Set $\epsilon = \eta \theta / \gamma$, and define $U_0 = B_\delta^e(u_0)$. From the differential equations above, it follows that for $\theta \neq 0$ the following identities hold:

$$u(t; u, X) = u(t\theta, u, X/\theta), \qquad v(t; u, X) = \theta v(t\theta; u, X/\theta).$$

Now $|t| < 2 \Leftrightarrow |t\theta| < 2\theta$ and, if $g_u(X, X) < \epsilon^2$,

$$(X/\theta) \cdot (X/\theta) \leq \gamma^2 g_u(X/\theta, X/\theta) \leq \gamma^2 \epsilon^2 / \theta^2 = \eta^2.$$

Therefore Φ is defined and differentiable on $\,]-2, 2[\, \times B_\epsilon U_0$, where

$$B_\epsilon U_0 = \{X \in TU_0 \mid g_{\pi(X)}(X, X) < \epsilon^2\}.$$

4. Let u_X be the representative of c_X in (U, g). Suppose $X \in B_\epsilon U_0$. Since $u_X(t) = u(t, \pi X, X)$, the map $X \to u_X(1)$ is differentiable. Since $u_X(t) = u(t, \pi X, X) = u(1, \pi X, tX) = u_{tX}(1)$, the set $\{tX \mid 0 \leq t \leq 1\}$ is mapped onto $\{u_X(t) \mid 0 \leq t \leq 1\}$. $\square$

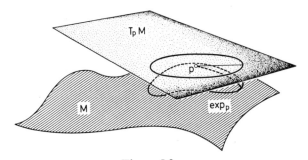

Figure 5.2

5.2.3 Definition. The map

$$B_\epsilon M_0 \to M; \qquad X \mapsto c_X(1)$$

is called the *exponential map* and is usually denoted by "exp." The open

5 Two-Dimensional Riemannian Geometry

sets M_0 and $B_\epsilon M_0$ corresponding to a given point $p_0 \in M$ are defined in (5.2.2), where the existence and smoothness of exp is proved.

Remark. The name exponential mapping comes from the theory of Lie groups. In the simplest possible case, the map $T_1 \mathbb{R}^+ \to \mathbb{R}^+$ given by $t \mapsto e^t$ is a map from the tangent space $T_1 \mathbb{R}^+$ of the multiplicative group $\mathbb{R}^+$ of positive reals (which we may identify with the additive group $\mathbb{R}$) into $\mathbb{R}^+$.

5.2.4 Lemma. *Let M be a surface with Riemannian metric. Let $p_0 \in M$. Then p_0 has a neighborhood $M_0 \subset M$ such that, for some $\epsilon > 0$, the map*

$$\pi \times \exp: B_\epsilon M_0 \to M \times M \quad \text{defined by } X \mapsto (\pi X, \exp X)$$

is an injective diffeomorphism (in other words, a diffeomorphism from $B_\epsilon M_0$ onto an open subset of $M \times M$).

PROOF. First translate the claim into a statement for a local coordinate system (U, g). Let $u_0 \in U$ be a representative of p_0. The map $\pi \times \exp: B_\epsilon U_0 \to U \times U$ exists and is differentiable by (5.2.2). Using (0.5.1), it will suffice to show that the differential of $\pi \times \exp$ is injective at $(u_0, 0)$. Toward that end, consider the curve $(u_0 + tX_0, tX)$ in $B_\epsilon U_0$. This curve passes through $(u_0, 0)$ when $t = 0$. What is its image in $M \times M$ under $\pi \times \exp$? Using the notation of (5.2.2), we see that it is $(u_0 + tX_0, u(t; u_0 + tX_0, X))$. This is because $u(1; u_0 + tX_0, tX) = u(t; u_0 + tX_0, X)$. Thus $d(\pi \times \exp)_{(u_0, 0)}(X_0, X) = (X_0, X + X_0)$, and therefore $d(\pi \times \exp)_{(u_0, 0)}$ is injective. □

As an easy corollary of this lemma, we have the following:

5.2.5 Theorem. *Let p_0 be a point on M, a surface with a Riemannian metric. Then there exists a neighborhood $M_0 \subset M$ of p_0 and a $\rho = \rho(p_0) > 0$ such that:*
  i) *Any two points $q, r \in M_0$ may be joined by a unique geodesic $c_{qr} = c_{qr}(t)$, $0 \leq t \leq 1$, of length $< \rho$.*
  ii) *The map $M_0 \times M_0 \to TM$ given by $(q, r) \mapsto \dot c_{qr}(0)$ is differentiable.*
  iii) *For every $q \in M_0$ the map $\exp_q: B_\rho(0) \subset T_q M \to M$ is an injective diffeomorphism (a diffeo onto an open subset of M).*

PROOF. Let (U, g) be a coordinate system for M. Let $u_0 \in U$. Choose ρ and $U_0' \ni u_0$ as in (5.2.4), making

$$\pi \times \exp: B_\rho U_0' \to U \times U$$

an injective diffeomorphism. Choose U_0 containing u_0 small enough so that $(\pi \times \exp)(B_\rho U_0') \supset U_0 \times U_0$. Therefore

$$(\pi \times \exp)^{-1}: U_0 \times U_0 \to B_\rho U_0'$$

is an injective diffeomorphism. What does this mean? Given v and w in U_0, $(\pi \times \exp)^{-1}(v, w) = X \in T_v \mathbb{R}^2$ is a tangent vector and $u_X(t)$, $0 \leq t \leq 1$, represents a geodesic of length $= |X| < \rho$. Moreover, $c_X(t)$ joins v to w. This proves (i) and (ii). (Why is $c_X(t)$ "unique"?)

Since $\exp_v B_\rho(0) = \text{pr} \circ (\pi \times \exp | B_\rho U_0' \cap T_v \mathbb{R}^2)$, where pr: $U \times U \to U$ is projection onto the second factor, (iii) follows. □

5.2.6 Definition. Let M be a surface with Riemannian metric. Suppose $\rho > 0$ is such that $\exp_p$ restricted to $B_\rho(0) \subset T_pM$ is an injective diffeomorphism from $B_\rho(0)$ into M. Then the image of $B_\rho(0)$ is called the ρ-*disc* with center p. It is denoted by $B_\rho(p)$.

The set $B_\rho(p) = \exp_p B_\rho(0)$ consists of precisely those points in M which may be joined to p by a geodesic of length less than ρ. (We know that every point in $B_\rho(p)$ may be joined to p by a geodesic of length $< \rho$. The converse follows from (5.3.4), below.)

5.3 Geodesic Polar Coordinates

5.3.1 Definition. Let M be a surface with a Riemannian metric. Let $p \in M$ be a point in M and let $\rho > 0$ be such that $B_\rho(p)$ is a ρ-disk with center at p. Let $\{e_1(p), e_2(p)\}$ be an orthonormal basis of T_pM.

i) The coordinate system $\phi: B_\rho(0) \subset T_pM \equiv \mathbb{R}^2 \to B_\rho(p)$, defined by $(v^1, v^2) \mapsto \exp_p(\sum_i v^i e_i(p))$, is known as *(Riemannian) normal coordinates*.

ii) *Geodesic polar* (or simply *polar*) *coordinates* on $B_\rho(p)$ are the coordinates

$$\phi:]0, \rho[\times \mathbb{R} \to B_\rho(p) - \{p\}: (r, \theta) \mapsto \exp_p(r\cos\theta e_1(p) + r\sin\theta e_2(p)).$$

The curves $r = $ constant are called *geodesic circles centered at p*.

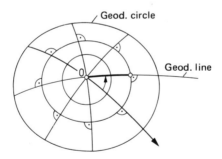

Figure 5.3   Geodesic polar coordinates

Remarks. i) Riemannian normal coordinates may be defined with respect to coordinate system (U, g) as follows. Let $u_0 \in U$ be a representative of p. Choose $\{e_1(u_0), e_2(u_0)\}$ an orthonormal basis of $T_{u_0}\mathbb{R}^2$ with respect to the metric g_{u_0}. Define $\phi: B_\rho(0) \to B_\rho(u_0)$ by $(v^1, v^2) \mapsto \exp_{u_0}(v^1 e_1 + v^2 e_2)$. Clearly ϕ is a diffeomorphism. Let $\tilde{g}(\_, \_) := g(d\phi, d\phi)$ be the induced metric on $B_\rho(0)$. Then $(B_\rho(0), \tilde{g})$ is a Riemannian coordinate system for $B_\rho(0)$.

ii) In order to make geodesic polar coordinates into a coordinate system in the usual sense, the θ variable must be restricted to lie in an open interval of length $< 2\pi$. For example,

$$\phi:]0, \rho[\times]-\pi, \pi[\to B_\rho(u_0) - \{-\rho te_1; 0 \le t < 1\}.$$

We have to remove an entire radius.

5.3.2 Proposition (Gauss–Lemma).[2] *Polar coordinates are geodesic coordinates based on a geodesic circle.*

PROOF. Let (U, g) be a coordinate system on M. As in the remark above, we may define

$$\phi: V :=]-\rho, \rho[\times]-\pi, \pi[\to B_\rho(u_0)$$
$$(v^1, v^2) \mapsto \exp_{u_0}(v^1 \cos v^2 e_1 + v^1 \sin v^2 e_2) =: (u^1, u^2).$$

We shall show that this is a geodesic coordinate system when $v^1 > 0$. To do this we shall use (4.3.6 (iii)), which means we must show that in these coordinates $\tilde{g}_{11} = 1$, $\tilde{g}_{12} = 0$, and $\tilde{g}_{22} > 0$. Now consider

$$\tilde{g}_{ij} = \sum_{k,l} g_{kl} \frac{\partial u^k}{\partial v^i} \frac{\partial u^l}{\partial v^j}$$

on V. Since $u^k(0, v^2) = u_0^k =$ constant for $k = 1, 2$, $\tilde{g}_{12}(0, v^2) = 0$. Fixing $v^2 = v_0^2$ and letting $v^1 = t \in]-\rho, \rho[$ vary parameterizes a unit-speed geodesic. Therefore $\tilde{g}_{11} = 1$ and, for $v^1 > 0$, $\tilde{\Gamma}^1_{11} = \tilde{\Gamma}^2_{11} = 0$. By definition of $\tilde{\Gamma}^2_{11}$,

$$\sum_i \tilde{g}^{2i}(2\tilde{g}_{i1,1} - \tilde{g}_{11,i}) = 2\tilde{g}^{22}\tilde{g}_{21,1} = 0.$$

But $\tilde{g}^{22} = \tilde{g}_{22}/\det(\tilde{g}_{ik}) \ne 0$, which means that $\tilde{g}_{21,1} = 0$ for $v^1 > 0$ and therefore for $v^1 \ge 0$ by continuity. $\tilde{g}_{12}(0, v^2) = 0$ implies that $\tilde{g}_{12} \equiv 0$. □

Our first application of the fact that geodesic polar coordinates are geodesic coordinates will be to show that geodesics have length minimizing properties analogous to those of straight lines in the plane, at least locally.

5.3.3 Definitions. On a surface M with Riemannian metric,
i) a curve $c = c(t)$, $t_0 \le t \le t_1$, from $p_0 = c(t_0)$ to $q = c(t_1)$ is *minimizing* if, for any curve $b = b(s)$, $s_0 \le s \le s_1$, from $p_0 = b(s_0)$ to $p_1 = b(s_1)$, $L(b) \ge L(c)$;
ii) a curve $c = c(t)$, $t \in I$, on M is *locally minimizing* if, for every $t_0 \in \mathring{I}$, there exists a closed interval $I_0 \subset I$ containing t_0 as interior point and on which c/I_0 is minimizing.

5.3.4 Theorem. *Let $B_\rho(p)$ be a ρ-disk centered at $p \in M$.*
i) *For every $q \in B_\rho(p)$, the geodesic $c = c_{pq} = c(t)$, $0 \le t \le 1$, defined in (5.2.5), is minimizing.*

---
[2] See footnote 12 of Chapter 6.

5.3 Geodesic Polar Coordinates

ii) If $b = b(s)$, $s_0 \leq s \leq s_1$, is any other curve from $p = b(s_0)$ to $q = b(s_1)$, then $L(b) \geq L(c)$ with equality if and only if there exists a diffeomorphism $t: [s_0, s_1] \to [0, 1]$ with $dt/ds \geq 0$ and $b(s) = c(t(s))$.

PROOF. 1. Without loss of generality $q \neq p$ and $L(c) = r_0 > 0$.

2. We may further assume that given any comparison curve $b(s)$, $s_0 \leq s \leq s_1$, then $b(s) \neq p$ for $s > s_0$. Introduce geodesic polar coordinates (5.3.1 (ii)) on $B_\rho(p) - \{p\}$. Here $(r, \theta) \in]0, \rho[\times \mathbb{R}$, and we may arrange it so that $\theta(c(t)) = 0$.

3. Suppose $b(s) \in B_\rho(p)$ for all $s \in [s_0, s_1]$. As in (2.1.3) one proves the existence of differentiable functions $\theta: [s_0, s_1] \to \mathbb{R}$ and $r: [s_0, s_1] \to]0, \rho[$ such that

$$b(s) = \exp_p(r(s) \cos \theta(s) e_1 + r(s) \sin \theta(s) e_2)$$

(this may also be proved directly). It follows that for $\epsilon > 0$ sufficiently small,

$$L(b \mid [s_0 + \epsilon, s_1]) = \int_{s_0+\epsilon}^{s} \sqrt{r'(s)^2 + g_{22}\theta'(s)^2}\, ds \geq r(s_1) - r(s_0 + \epsilon)$$
$$= L(c) - r(s_0 + \epsilon).$$

Since $r(s_0 + \epsilon) \to 0$ as $\epsilon \to 0$, $L(b) \geq L(c)$.

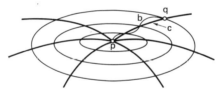

Figure 5.4 Geodesics are locally minimizing

4. Suppose $b(s)$ leaves the set $B_\rho(p)$. This means that there exists an $s_2 < s_1$ such that $b \mid [s_0, s_2] \subset B_\rho(p)$ and $L(c) < r(s_2) < \rho$. Therefore $L(b) \geq r(s_2) > L(c)$.

5. Suppose $L(b) = L(c)$. Looking at the inequality in (3), we see that the only way to get equality is for $\theta'(s) \equiv 0$ and $r'(s) \geq 0$. Therefore $\theta(s) \equiv 0$. Letting $t(s) = r(s)/r_0$, where $r_0 = L(c)$, produces the required change of parameter. □

5.3.5 Theorem (A characterization of geodesics). *A curve $b = b(s)$, $s_0 \leq s \leq s_1$, on M is locally minimizing if and only if there exists a smooth mapping $t: [s_0, s_1] \mapsto [0, 1]$ with $dt/ds \geq 0$ such that $b(s) = c(t(s))$, where c is a geodesic.*

PROOF. By (5.3.4), b is locally minimizing implies that b is locally of the form $c(t(s))$. Conversely, (5.2.5) and (5.3.4) together show that geodesics are locally minimizing, since length remains unchanged under a change of parameters $s \mapsto t(s)$ with $dt/ds \geq 0$. □

5.4 Jacobi Fields

5.4.1 Definition. Let $c = c(t)$, $0 \leq t \leq a$, be a unit-speed geodesic on M. A vector field $Y(t)$ along c is a *Jacobi field* provided $g_{c(t)}(\dot{c}(t), Y(t)) = 0$, i.e., Y is orthogonal to c, and

(*) $$\frac{\nabla^2 Y}{dt^2}(t) + K \circ c(t) Y(t) = 0.$$

This definition is clearly coordinate invariant, i.e., independent of the choice of a coordinate system (U, g) on M. It will be useful to have (*) expressed in terms of the Frenet frame $e_1(t), e_2(t)$ on c. We may write $Y(t) = y_2(t) e_2(t)$ for some smooth function $y(t)$. Then (*) is equivalent to

$$\ddot{y}(t) + K \circ c(t) y(t) = 0.$$

This follows since $\nabla^i e_2(t)/dt^i = 0$. As a further consequence of this,

$$g_{c(t)}\left(\dot{c}(t), \frac{\nabla Y(t)}{dt}\right) = g_{c(t)}\left(e_1(t), \dot{y}(t) e_2(t) + y(t) \frac{\nabla e_2(t)}{dt}\right) = 0.$$

5.4.2 Proposition. *Let $c(t)$ be a unit-speed geodesic ($|\dot{c}(t)| = 1$). Given $a_0, a_1 \in \mathbb{R}$, there exists a unique Jacobi field $Y(t) = y(t) e_2(t)$ with $y(0) = a_0$, $\dot{y}(0) = a_1$.* □

This follows directly from the existence and uniqueness theorem for ordinary differential equations.

5.4.3 Lemma (How to produce a Jacobi field). *Let $c(t)$, $0 \leq t \leq a$, be a geodesic with $|\dot{c}(t)| = 1$ and call $c(0) = p$. Let $\tilde{c}(t)$ denote the segment $t\dot{c}(0)$, $0 \leq t \leq a$, in the tangent space $T_p M$. Let $A \in T_p M$ be a vector orthogonal to $\dot{c}(0)$. Then*

$$Y(t) := (d \exp_p)_{\tilde{c}(t)}(tA) \in T_{c(t)} M$$

is a Jacobi field along $c(t)$. Moreover, $Y(t)$ satisfies the initial conditions $Y(0) = 0$ and $\nabla Y/dt(0) = A$. (Here we consider tA as an element of $T_{\tilde{c}(t)} M_p$ via the canonical identification.) Since a Jacobi field $Y(t)$ is completely determined by the initial conditions $Y(0)$, $(\nabla Y/dt)(0)$, every Jacobi field $Y(t)$ with $Y(0) = 0$ may be written in the above form.

PROOF. 1. Without loss of generality we may assume that $A \neq 0$. Furthermore, solutions to the Jacobi equation (*) form a vector space; in particular, if $Y(t)$ is a Jacobi field so is $a \cdot Y(t)$, $a \in \mathbb{R}$. Therefore we may assume that $|A| = 1$.

2. Consider the orthonormal basis $\{e_1(p), e_2(p)\} = \{\dot{c}(0), A\}$ in $T_p M$. For sufficiently small $\delta, \epsilon > 0$, we will define $\phi: [0, a + \delta[\times]-\epsilon, \epsilon[\to M$ by

$$(r, \theta) \mapsto \exp_p(r \cos \theta\, e_1(p) + r \sin \theta e_2(p)).$$

5.4 Jacobi Fields

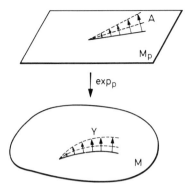

Figure 5.5 Generation of a Jacobi field by variation through geodesics

For sufficiently small $r > 0$, this is a polar coordinate system centered at p. We need to show that there exists $\delta > 0$ and $\epsilon > 0$ so that ϕ is defined. Notice that $\phi(t, 0)$ is defined for $t \in [0, a]$; in fact, $\phi(t, 0) = c(t)$. Moreover, $\phi([0, a], 0)$ is compact. If $U \subset T_p M$ is the domain of definition of $\exp_p$, then $\phi([0, a], 0)$ lies in $\exp_p(U)$, an open set. The existence of the required $\epsilon > 0$ and $\delta > 0$ now follow from the compactness of $\phi([0, a], 0)$.

Let $\{e_1(t), e_2(t)\}$ be the Frenet frame along $c(t)$ with $\{e_1(0), e_2(0)\} = \{e_1(p), e_2(p)\}$. We consider the (t, θ) coordinate having coordinate basis $\{e_1(t, \theta), e_2(t, \theta)\}$. Now $Y(t) = (\partial \phi / \partial \theta)(t, 0) = d\phi(e_2(t, 0))$, so if we write $Y(t) = y(t)e_2(t)$, then $y(t)^2 = |Y(t)|^2$. Wherever ϕ is a coordinate system, its first fundamental form (g_{ij}) must have $g_{22}(t, 0) = y(t)^2$. In fact, ϕ will be a coordinate system in a neighborhood of any point $(t, 0)$ where $y(t) \neq 0$, i.e., where $Y(t) \neq 0$. On such a neighborhood, we have geodesic polar coordinates and hence, by (5.3.2), geodesic coordinates. This allows us to use the formula for Gauss curvature:

$$K = \frac{-(\sqrt{g_{22}})_{,11}}{g_{22}}$$

of (4.3.8), where we consider $(t, \theta) = (u^1, u^2)$. Since $g_{22}(t, 0) = y(t)^2$, the above formula implies that at least for $t \in \mathring{I}$ where $y(t) \neq 0$,

(*) $$\ddot{y}(t) + K \circ c(t)y(t) = 0.$$

How do we handle the points where $y(t) = Y(t) = 0$? Such points must be isolated: for if t_0 were a nonisolated point of this set, then $Y(t_0)$ would be the unique Jacobi field with $Y(t_0) = \nabla Y(t_0)/dt = 0$, i.e., $Y(t) \equiv 0$, contradicting the fact that $\nabla Y(0)/dt = A \neq 0$. Now $y(t)$ is defined and differentiable for all t and satisfies (*) except at isolated points. It follows by continuity of $y(t)$ that $y(t)$ satisfies the equation (*) everywhere.

3. We now calculate $(\nabla Y/dt)(0)$. Letting $(t, \theta) = (u^1, u^2)$, the geodesic

103

5 Two-Dimensional Riemannian Geometry

$c(t)$ is representable as $\phi \circ u$, where $u^1(t) = t$, $u^2(t) = 0$, and $Y(t) = d\phi(e_2(u^1(t), u^2(t))) = d\phi(t(\partial/\partial u^2)) = t(\partial\phi/\partial u^2)(t, 0)$. Using (4.1.2),

$$\frac{\nabla Y}{dt}(0) = \lim_{t\to 0} \frac{\nabla Y}{dt}(t) = \lim_{t\to 0}\left(e_2(t, 0) + \sum_k t\Gamma^k_{21}e_k(t, 0)\right) = e_2(0)$$

$$= e_2(p) = A. \qquad \square$$

Remark. It follows from (3) of the proof that in geodesic polar coordinates, $(u^1, u^2) = (t, \theta) \mapsto \exp_p(t\cos\theta e_1 + t\sin\theta e_2)$, $\sqrt{g_{22}(t, \theta_0)}$ is equal to the length of the unique Jacobi field $Y(t)$ along $\gamma_{\theta_0}(t) = \exp_p(t\cos\theta_0 e_1 + t\sin\theta_0 e_2)$ with $Y(0) = 0$ and $(\nabla Y/dt)(0) = -\sin\theta_0 e_1 + \cos\theta_0 e_2$.

5.4.4 Proposition. *Let $Y(t) = y(t)e_2(t)$ be a Jacobi field along $c(t)$ with $y(0) = 0$, $\dot{y}(0) = 1$. Then we have the following Taylor series expansion for $y(t)$ at $t = 0$:*

$$y(t) = t - K \circ c(0) \cdot \frac{t^3}{6} + \cdots.$$

PROOF. Immediate from the differential equation

$$\ddot{y} + K \circ c(t) \cdot y(t) = 0. \qquad \square$$

We now use this proposition to prove several interesting results about the geometry of M near p. We assume $B_\rho(p)$ is an embedded geodesic disk.

5.4.5 Proposition. i) *Let $L(r)$ be the length of a geodesic circle $S^1_r(p)$ of radius r in $B_\rho(p)$. Then we have the following Taylor expansion for $L(r)$ at $r = 0$:*

$$L(r) = 2\pi r - 2\pi K(p) \cdot \frac{r^3}{6} + \cdots.$$

ii) *Let $A(r)$ be the area of the r-disk $B_r(p)$ centered at p, $r \leq \rho$. Then we have the following Taylor expansion for $A(r)$ at $r = 0$:*

$$A(r) = \pi r^2 - \pi K(p) \cdot \frac{r^4}{12} + \cdots.$$

Remark. The notion of the area of a subset of M is defined in (5.6.6).

As an immediate corollary of (5.4.5) we get a striking theorem which relates Gauss curvature to the deviation of the geometric functions $L(r)$ and $A(r)$ on a surface M from the corresponding Euclidean quantities.

5.4.6 Theorem. *Let $L(r)$ and $A(r)$ be defined as in (5.4.5). Then*

$$K(p) = \lim_{r\to 0} \frac{2\pi r - L(r)}{r^3} \cdot \frac{3}{\pi} = \lim_{r\to 0} \frac{\pi r^2 - A(r)}{r^4} \cdot \frac{12}{\pi}.$$

PROOF of (5.4.5). Let $\{e_1(p), e_2(p)\}$ be an orthonormal basis of T_pM. The unit circle in T_pM is $\tilde{b}(s) = \cos s \cdot e_1(p) + \sin s \cdot e_2(p)$, and the geodesic circles $S_r^1(p)$ may be expressed as $c(s) = \exp_p r\tilde{b}(s)$, $0 \le s \le 2\pi$. Using (5.4.3) we may interpret $c'(s) = (d \exp_p)_{r\tilde{b}(s)} r\tilde{b}'(s)$ as the value at $t = r$ of the Jacobi field $Y(t; s) = (d \exp_p)_{t\tilde{b}(s)} t\tilde{b}'(s)$ along $\exp_p t\tilde{b}(s)$, $0 \le t \le r$. Since $|\tilde{b}'(s)| = 1$, (5.4.4) implies that $|Y(r; s)| = r - K(p) \cdot r^3/6 + \ldots$ for r small. Therefore

$$L(S_r^1(p)) = \int_0^{2\pi} |Y(r, s)| \, ds = \int_0^{2\pi} \left(r - K(p) \cdot \frac{r^3}{6} + \ldots \right) ds$$

and

$$A(B_r(p)) = \int_0^r \int_0^{2\pi} |Y(t, s)| \, ds \, dt = \int_0^r \int_0^{2\pi} \left(t - K(p) \cdot \frac{t^3}{6} + \ldots \right) ds \, dt,$$

which proves the proposition. □

5.5 Manifolds

We will now introduce the second generalization of the idea of a surface. Up to now, we have required a surface to be representable in terms of one single coordinate system (U, g). This restriction will now be dropped. It will now be possible to treat, for example, the entire sphere S^2 in $\mathbb{R}^3$ as a surface. Until now, we have had to consider only a part of S^2, e.g., S^2 minus half of a great circle as in (3.3.7).

Furthermore, it will be useful to allow our generalized surfaces to have arbitrary dimension, and not restrict them to dimension 2. We have already seen that investigating surfaces of dimension 2 leads to the introduction of the tangent bundle, a four-dimensional object.

5.5.1 Definitions. i) A *topological manifold* M of dimension n is a Hausdorff topological space with a countable basis such that there exists a family of homeomorphisms $\{u_\alpha : M_\alpha \to U_\alpha \subset \mathbb{R}^n\}_{\alpha \in A}$ from open sets $M_\alpha \subset M$ to open sets $U_\alpha \subset \mathbb{R}^n$ and $\bigcup_\alpha M_\alpha = M$. These homeomorphisms will usually be denoted by (u_α, M_α), and they are called *coordinate systems* or *charts* for M. The collection $(u_\alpha, M_\alpha)_{\alpha \in A}$ is called a *(topological) atlas* for M.

ii) An atlas $(u_\alpha, M_\alpha)_{\alpha \in A}$ is a *differentiable atlas* if, for every $(\alpha, \beta) \in A \times A$, the homeomorphism $u_\beta \circ (u_\alpha | M_\alpha \cap M_\beta)^{-1} : u_\alpha(M_\alpha \cap M_\beta) \to u_\beta(M_\beta \cap M_\alpha)$ is a diffeomorphism.

iii) Two atlases $(u_\alpha, M_\alpha)_{\alpha \in A}$ and $(u_{\alpha'}, M_{\alpha'})_{\alpha' \in A'}$ are *equivalent* if the union of these atlases is a differentiable atlas.

iv) A *differentiable manifold* is a topological manifold together with an equivalence class of differentiable atlases.

Remark. For the case $n = \dim M = 2$, a manifold is also called a surface. These will be the focus of our study.

5 Two-Dimensional Riemannian Geometry

It is clear that (iii) defines an equivalence relation. For two equivalent atlases $(u_\alpha, M_\alpha)_{\alpha \in A}$ and $(u_{\alpha'}, M_{\alpha'})_{\alpha' \in A'}$ every $u_{\alpha'} \circ u_\alpha^{-1}$, $(\alpha, \alpha') \in A \times A'$ is a diffeomorphism. *Note: From now on, when we speak of an atlas we will always mean a differentiable atlas.*

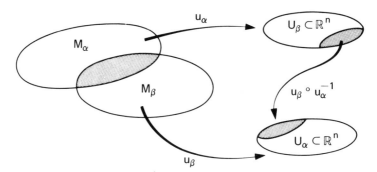

Figure 5.6 Change of coordinates

The concept of differentiable manifold allows us to define what it means for a function $F: M \to N$ between differentiable manifolds to be a differentiable function.

5.5.2 Definition. Suppose M and N are differentiable manifolds and $F: M \to N$ is a continuous function. Then F is differentiable if, for atlases $(u_\alpha, M_\alpha)_{\alpha \in A}$ of M and $(v_\beta, N_\beta)_{\beta \in B}$ of N, the function

$$v_\beta \circ F \circ u_\alpha^{-1}: u_\alpha(M_\alpha \cap F^{-1}(N_\beta)) \to v_\beta(N_\beta)$$

is differentiable for all $(\alpha, \beta) \in A \times B$.

This definition is independent of the choice of atlases as one may readily see from the equality

$$v_{\beta'} \circ F \circ u_{\alpha'}^{-1} = (v_{\beta'} \circ v_\beta^{-1}) \circ (v_\beta \circ F \circ u_\alpha^{-1}) \circ (u_\alpha \circ u_{\alpha'}^{-1}).$$

EXAMPLE. A curve $c: I \to M$ is differentiable provided: for every chart (u_α, M_α), $c \mid (I \cap c^{-1}(M_\alpha))$ is represented by a differentiable function $C_\alpha: t \in I \cap c^{-1}(M_\alpha) \to u_\alpha(M_\alpha)$. We consider I as a one-dimensional differentiable manifold with atlas consisting of the single chart (id, I).

5.5.3 Some examples of (differentiable) surfaces and manifolds

1. *The sphere* $M = S_r^2(0) = \{(x, y, z) \in \mathbb{R}^3 \mid x^2 + y^2 + z^2 = r^2\}$ with the topology induced from $\mathbb{R}^3$. Since it is a subset of $\mathbb{R}^3$, M is Hausdorff and has a countable basis of open sets. We may define an atlas consisting of two charts, $\{u_+, M_+\}, \{u_-, M_-\}$, as follows:

$$M_+ = M - \{(0, 0, -r)\}; \qquad M_- = M - \{(0, 0, r)\}$$

$$u_+(x, y, z) = \left(\frac{rx}{(r+z)}, \frac{ry}{(r+z)} \right)$$

$$u_-(x, y, z) = \left(\frac{rx}{(r-z)}, \frac{ry}{(r-z)} \right) =: (\xi, \eta).$$

The maps u_+ and u_- are stereographic projections from the south and north poles, respectively.

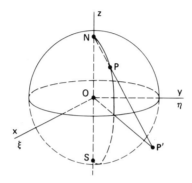

Figure 5.7 Stereographic projection from North Pole; $u_-(P) = P'$

Let $(x, y, z) = U_-^{-1}(\chi, \eta)$. Then

$$x = \frac{2\xi r^2}{(\xi^2 + \eta^2 + r^2)}, \quad y = \frac{2\eta r^2}{(\xi^2 + \eta^2 + r^2)}, \quad z = \frac{r(\xi^2 + \eta^2 - r^2)}{(\xi^2 + \eta^2 + r^2)},$$

and the map

$$u_+ \circ u_-^{-1}(\xi, \eta) = \left(\frac{r^2 \xi}{(\xi^2 + \eta^2)}, \frac{r^2 \eta}{(\xi^2 + \eta^2)} \right)$$

is a diffeomorphism of $u_-(M_- \cap M_+) = \mathbb{R}^2 - \{(0, 0)\}$ onto $u_+(M_+ \cap M_-) = \mathbb{R}^2 - \{(0, 0)\}$. It is easy to see that $\det (d(u_+ \circ u_-^{-1})) < 0$.

2. *The projective plane* $M = P^2$. Consider the set

$$P^2 := \{\{x, -x\} \mid x \in \mathbb{R}^3, |x| = 1\}.$$

We define a topology on M as follows. Let $S^2 := \{x \in \mathbb{R}^3 \mid |x| = 1\} = S_1^2(0)$. Consider the mapping $\varphi: S^2 \to P^2$ given by $x \mapsto \{x, -x\}$. If $B \subset S^2$ lies in an open hemisphere, $\varphi \mid B: B \to P^2$ is injective. As a basis for the topology of P^2 we will take the collection of sets $\varphi(B)$, where $B \subset S^2$ lies in an open hemisphere of S^2.

If $(u_\alpha, S_\alpha^2)_{\alpha \in A}$ is a differentiable atlas for S^2 which has the property that each S_α^2 lies in an open hemisphere, $u_\alpha \circ (\varphi \mid S_\alpha^2)^{-1}$, $(\varphi(S_\alpha^2))_{\alpha \in A}$ is a differentiable

5 Two-Dimensional Riemannian Geometry

atlas for P^2. For example, we may take the atlas $(u_\pm, M_\pm)$ of S^2 which was developed in (1) above and subdivide it to give an atlas of S^2 with the required property. Thus P^2 is a differentiable surface. Also, our construction makes $\varphi: S^2 \to P^2$ a differentiable mapping.

We certainly expect differentiable manifolds to have all the general properties that locally defined surfaces have. In particular, the notion of a tangent vector should be a natural one. In the interest of clarity, we will restrict ourselves to the case of surfaces. The general case (arbitrary dimension) may be treated in the same manner.

5.5.4 Definition. Let M and N be differentiable surfaces.
  i) Suppose (u_α, M_α) is a chart for M and $p \in M_\alpha$. The vector space $T_{u_\alpha(p)}\mathbb{R}^2$ is a *representation of the tangent space of M at p*. A vector $X_\alpha \in T_{u_\alpha(p)}\mathbb{R}^2$ is a *representative of a tangent vector to M at p*.
  ii) Suppose (u_α, M_α) and (u_β, M_β) are two charts on M and $p \in M_\alpha \cap M_\beta$. The representatives $X_\alpha \in T_{u_\alpha(p)}\mathbb{R}^2$ and $X_\beta \in T_{u_\beta(p)}\mathbb{R}^2$ are *equivalent* (or *represent the same tangent vector*) provided $X_\beta = d(u_\beta \circ u_\alpha^{-1})X_\alpha$.
  iii) The equivalence of (ii) is an equivalence relation. An equivalence class of vectors is called a *tangent vector to M at p*.
  iv) The set of tangent vectors to M at p carries a vector space structure determined by the vector space structure of any one of its representations. This vector space is called the *tangent space to M at p*, and will be denoted by $T_p M$.

Remark. These definitions are compatible with the corresponding definitions of tangent vector and tangent space for surfaces (U, g), (5.1.1). We will now define the tangent bundle of a differentiable manifold. Toward that end, we first show that $\bigcup_{p \in M} T_p M$ has a naturally defined differentiable atlas.

5.5.5 Proposition. *Let M be a differentiable surface. Let $TM := \bigcup_{p \in M} T_p M$ denote the union of all tangent spaces to points $p \in M$. Let $\pi: TM \to M$ be the projection mapping $X \mapsto p$ when $X \in T_p M$. Then TM is a four-dimensional differentiable manifold whose differentiable structure is determined by that of M: Given $(u_\alpha, M_\alpha)_{\alpha \in A}$, an atlas for TM is defined by $(Tu_\alpha, TM_\alpha)_{\alpha \in A}$, where $TM_\alpha = \bigcup_{p \in M_\alpha} T_p M$ and $Tu_\alpha: TM_\alpha \to TU_\alpha$ is the map $X \in T_p M \mapsto X_\alpha \in T_{u_\alpha(p)}\mathbb{R}^2$, where X_α is a representative of X.*

With this differentiable structure on TM, the projection $\pi: TM \to M$ is differentiable.

PROOF. 1. First we define a topology for TM as follows. The map $Tu_\alpha: TM_\alpha \to TU_\alpha$ is bijective and its image is an open subset of $\mathbb{R}^2 \times \mathbb{R}^2$. A set S in TM_α is open if and only if $Tu_\alpha(S)$ is open in $\mathbb{R}^2 \times \mathbb{R}^2$.
2. The bijection

$$Tu_\beta \circ Tu_\alpha^{-1}: Tu_\alpha(T(M_\alpha \cap M_\beta)) \to Tu_\beta(T(M_\beta \cap M_\alpha))$$

is in the form $d(u_\beta \circ u_\alpha^{-1})$ and is thus a diffeomorphism. Therefore the topology on TM_α is independent of the choice of coordinates. Furthermore, this implies that $(Tu_\alpha, TM_\alpha)_{\alpha \in A}$ is a differentiable atlas.
3. If $(u_{\alpha'}, M_{\alpha'})_{\alpha' \in A'}$ is an atlas which is equivalent to $(u_\alpha, M_\alpha)_{\alpha \in A}$, then $(Tu_{\alpha'}, TM_{\alpha'})_{\alpha' \in A'}$ is an atlas equivalent to $(Tu_\alpha, TM_\alpha)_{\alpha \in A}$. This follows directly from the definitions and is easy to check. Therefore the differentiable structure of TM is determined by the differentiable structure of M.
4. It remains to show that $\pi: TM \to M$ is differentiable. This is a local question, so let us consider $\pi | TM_\alpha$. In terms of the coordinate chart (Tu_α, TM_α), $\pi_1 = u_\alpha \circ \pi \circ Tu_\alpha^{-1}$, where $\pi_1: TU_\alpha = U_\alpha \times \mathbb{R}^2 \to U_\alpha$ is the projection onto the first factor. It follows that π is differentiable. □

5.5.6 Definition. The *tangent bundle* TM of a differentiable surface M is the four-dimensional differentiable manifold defined in (5.5.5).

It is now possible to define vector fields on M.

5.5.7 Definition. A *vector field* on M is a differentiable mapping $X: M \to TM$ which satisfies $\pi \circ X = \text{id}$. In other words, $X(p) \in T_p M$.

Remark. A chart (u_α, M_α) of M defines two linearly independent vector fields on M_α, namely vector fields represented by the basis vector fields $e_1(u_\alpha)$, $e_2(u_\alpha)$ on TU_α. However, it is not always possible to find two linearly independent vector fields defined on all of M. In fact, it can be proved that if M is a compact surface, the existence of two globally defined linearly independent vector fields implies that M is a torus. That such vector fields do exist on a torus follows from (3.3.7 (ii)): The vector fields desired can be constructed by taking the tangent vectors to the globally defined parameter curves corresponding to the (u, v) coordinates.

5.5.8 Definition. A *surface with a Riemannian metric* is a differentiable surface such that, for each $p \in M$, $T_p M$ has a positive definite inner product which is a differentiable function of $p \in M$. In terms of an atlas $(u_\alpha, M_\alpha)_{\alpha \in A}$, this means that, for every $\alpha \in A$, there exists a $g_\alpha(\, , \,): U_\alpha \to S(2)$ such that given $(\alpha, \beta) \in A \times A$, $u_\beta \circ u_\alpha^{-1}: u_\alpha(M_\alpha \cap M_\beta) \mapsto u_\beta(M_\beta \cap M_\alpha)$ is an isometry.

A *manifold with a Riemannian metric* is defined analogously.

Remark. This definition includes two different ways to think about the metric on a surface. First, it may be conceived of as an inner product $g(\, , \,)$ on each $T_p M$ which in terms of a chart (u_α, M_α) corresponds to an inner product g_α on U_α. Using the notation of (5.5.5),

$$g(X, Y) = g_\alpha(Tu_\alpha X, Tu_\alpha Y) \quad \text{for } X, Y \in T_p M.$$

5 Two-Dimensional Riemannian Geometry

The requirement that the inner product be a differentiable function of p is equivalent to requiring each g_α to be differentiable. Equivalently, one could require that, given any two differentiable fields X, Y, the function $p \mapsto g(X(p), Y(p))$ be differentiable.

One may reverse the procedure and consider the Riemannian metric as being given by the collection $(U_\alpha, g_\alpha)_{\alpha \in A}$ corresponding to an atlas $(u_\alpha, M_\alpha)_{\alpha \in A}$ of M. The identification of TM_α with TU_α via Tu_α defines an inner product on each T_pM, $p \in M$. If $p \in M_\alpha \cap M_\beta$, there are two different ways to define an inner product on T_pM. The question is: Do they agree? The answer is yes if and only if $u_\beta \circ u_\alpha^{-1}: u_\alpha(M_\alpha \cap M_\beta) \to u_\beta(M_\alpha \cap M_\beta)$ is an isometry.

In sections (5.1) through (5.4) we considered surfaces with Riemannian metrics which were representable in terms of a single coordinate system (U, g). All of the concepts and definitions we introduced there as well as the theorems and propositions concerning them carry over word for word to surfaces (and manifolds!) with a Riemannian metric. For example, let $c: I \to M$ be a curve (see the example preceding (5.5.3)). The vectors $\dot{c}(t) \in T_{c(t)}M$ are well defined and therefore we may also define

$$L(c) := \int_I \sqrt{g_{c(t)}(\dot{c}(t), \dot{c}(t))}\, dt$$

$$E(c) := \frac{1}{2} \int_I g_{c(t)}(\dot{c}(t), \dot{c}(t))\, dt,$$

the length and energy of c.

To end this section we now define the concept of an *orientable surface*. (This concept is only of interest for surfaces (and manifolds) which cannot be represented in terms of a single coordinate system. If a surface consists of a single chart then it trivially satisfies the definition.)

5.5.9 Definition. Let M be a differentiable surface (or manifold).
  i) M is said to be *orientable* if there exists an atlas (u_α, M_α) with the following property:

$$\det(d(u_\beta \circ u_\alpha^{-1})) > 0 \quad \text{for all } (\alpha, \beta) \in A \times A.$$

  The atlas itself is also said to be *orientable*.
  ii) Two orientable atlases have the same *orientation* provided their union is orientable. This is an equivalence relation among orientable atlases on M. An equivalence class is also called an *orientation*.
  iii) An *oriented manifold* is a manifold together with a distinguished orientation, designated as *positive*. A chart belonging to one of the atlases is called a *positively oriented* chart.

EXAMPLES AND DISCUSSION. 1. If M is orientable and connected, there exist exactly two orientations of M (i.e., two equivalence classes of atlases under the equivalence relation in (ii). (Proof: exercise.)

2. Not every surface (or manifold) is orientable. For example, the projective plane P^2 defined in (5.5.3, 2) is *not* orientable. To see this, consider the *antipodal* map $i: S^2 \to S^2$ which maps $x \mapsto -x$. In terms of the atlas for S^2 defined in (5.5.3, 1),

$$u_+ \circ i \circ u_-^{-1}(\xi, \eta) = (-\xi, -\eta); \qquad (\xi, \eta) \neq (0, 0).$$

It follows that this map reverses orientation.* Now assume that P^2 possesses an oriented atlas $(u_\alpha, P_\alpha^2)_{\alpha \in A}$. Recall that $\varphi: S^2 \to P^2$ is the map $x \mapsto \{x, -x\}$. The sets $\varphi^{-1} P_\alpha^2$ can be divided into sets $S_\alpha^2 \cup i S_\alpha^2$ in such a way that $\varphi: S_\alpha^2 \to P^2$ and $\varphi: i S_\alpha^2 \to P^2$ are diffeomorphisms. Thus $(u_\alpha \circ \varphi, S_\alpha^2)_{\alpha \in A} \cup (u_\alpha \circ \varphi, i S_\alpha^2)_{\alpha \in A}$ is an orientable atlas for S^2. But $i: S_\alpha^2 \to i S_\alpha^2$ has the coordinate representation id: $u_\alpha(P_\alpha^2) \to u_\alpha(P_\alpha^2)$. This is a contradiction since i is orientation reversing, but id: $u_\alpha(P_\alpha^2) \to u_\alpha(P_\alpha^2)$ is not.

5.6 Differential Forms

5.6.1 More linear algebra. We continue the development of (3.2.1). Let T be a real vector space of dimension n. For our purposes, n will usually be equal to 2.

1. The *dual space* T^* of T is the set $\mathscr{L}(T, \mathbb{R})$ of linear mappings $\omega: T \to \mathbb{R}$, together with the natural vector space structure

$$(\omega_1 + \omega_2)(X) = \omega_1(X) + \omega_2(X), \qquad (a\omega)(X) = a\omega(X).$$

The elements of T^* are called 1-*forms* or *linear functionals* (on T). If e_i, $1 \leq i \leq n$, is a basis of T, we may define a basis e^j, $1 \leq j \leq n$, of T^*, the dual basis, by the equations

$$e^j(e_i) = \delta_i^j.$$

2. Let f_k, $1 \leq k \leq m$, be a basis for another vector space S. If $L: S \to T$ is a linear mapping, we may write $L f_k = \sum_i a_k^i e_i$ for some $n \times m$-matrix (a_k^i). The *dual mapping* $L^*: T^* \to S^*$ of L is the mapping defined by the relation $L^*(\omega) = \omega \circ L$. This implies that $L^*(e^j) = e^j \circ L = \sum_i a_i^j f^i$. (The matrix of L^* is the transpose of the matrix of L.) If L is bijective then L^* is also bijective and $L^{*-1}: S^* \to T^*$ may be written in terms of a basis as $L^{*-1} f^k = \sum_j b_j^k e^j$, where the matrix (b_j^k) satisfies $\sum_j b_j^k a_i^j = \delta_i^k$. (In other words, $(b_j^k) = ({}^t(a_j^k))^{-1}$.)

3. The *direct sum* $T \oplus T$ of T with itself is the set of pairs $(X, Y) \in T \times T$ with the vector space structure:

$$(X_1, Y_1) + (X_2, Y_2) = (X_1 + X_2, Y_1 + Y_2)$$
$$a(X, Y) = (aX, aY).$$

4. A 2-*form* on T is a mapping $\Omega: T \oplus T \to \mathbb{R}$ which is bilinear and skew-symmetric:
   i) $\Omega(aX + bY, Z) = a\Omega(X, Z) + b\Omega(Y, Z)$,

---
* If $j(\xi, \eta) = (\xi, -\eta)$, $\{(j \circ u_+, M_+), (u_-, M_-)\}$ is an orientable atlas for S^2.

ii) $\Omega(X, Y) = -\Omega(Y, X)$.
(*Note:* Linearity in the second variable follows from (ii) together with (i), where $X, Y, Z \in T$ and $a, b \in \mathbb{R}$.)

The set $\Lambda^2 T^*$ of all 2-forms on T is a vector space with the following addition and scalar multiplication:

$$(\Omega + \Omega')(X, Y) = \Omega(X, Y) + \Omega'(X, Y)$$
$$(a\Omega)(X, Y) = a\Omega(X, Y).$$

For $n = 2$ we will show that $\Lambda^2 T^*$ has dimension $= 1$. To wit, if e_1, e_2 is a basis for T, we define an element $e^1 \wedge e^2 \in \Lambda^2 T^*$ by

$$e^1 \wedge e^2(X, Y) = \xi^1 \eta^2 - \xi^2 \eta^1 = \det(\xi, \eta),$$

where $X = \sum_i \xi^i e_i$ and $Y = \sum_j \eta^j e_j$. If Ω is an arbitrary element of $\Lambda^2 T^*$,

$$\Omega(X, Y) = \sum_{i,j=1,2} \xi^i \eta^j \Omega(e_i, e_j)$$
$$= (\xi^1 \eta^2 - \xi^2 \eta^1)\Omega(e_1, e_2) = A(e^1 \wedge e^2)(X, Y),$$

where $A = \Omega(e_1, e_2)$. Therefore $e^1 \wedge e^2$ spans $\Lambda^2 T^*$.

5. Let $L: S \to T$ be a linear mapping as in (2) above. Then we may define the mapping $\Lambda^2 L^*: \Lambda^2 T^* \to \Lambda^2 S^*$ by $(\Lambda^2 L^* \Omega)(X, Y) = \Omega(LX, LY)$.

In the special case that $\dim S = \dim T = 2$,

$$(\Lambda^2 L^* \Omega)(f_1, f_2) = \Omega(Lf_1, Lf_2) = \Omega(e_1, e_2) \cdot \det(a_j^k),$$

where $Lf_k = \sum_i a_k^i e_i$. But by (3) this equation implies that $\Lambda^2 L^*(e^1 \wedge e^2) = \det(a_i^k) f^1 \wedge f^2$.

6. Suppose $\beta: T \times T \to \mathbb{R}$ is a quadratic form on T. Then β defines an associated linear mapping $L_\beta: T \to T^*$, namely

$$L_\beta: T \to T^*, \qquad X \to \beta(X, \ ).$$

Note that L_β is bijective if and only if, for every $X \neq 0$, there exists a Y such that $\beta(X, Y) \neq 0$. A quadratic form with the above property is called *nondegenerate*. For example, a positive definite quadratic form is nondegenerate. Another way to characterize a nondegenerate form is: β is nondegenerate if and only if its matrix representation has nonzero determinant. A very important fact about a nondegenerate quadratic form β is that we may use it to identify T with T^* by means of the bijection L_β.

We would now like to generalize the above definitions to surfaces. In order to do this we must, among other things, generalize the idea of a direct sum of vector spaces to its counterpart for tangent bundles.

5.6.2 Definition. Let M be a differentiable surface. The *direct sum* $TM \oplus TM$ of the tangent bundle of M with itself is the disjoint union $\bigcup_{p \in M} T_p M \oplus T_p M$. The projection mapping $\pi \oplus \pi: TM \oplus TM \to M$ is the mapping defined by

$$(X, Y) \in T_p M \oplus T_p M \mapsto p.$$

5.6 Differential Forms

5.6.3 Proposition. $TM \oplus TM$ can be given the structure of a six-dimensional differentiable manifold which we also denote by $TM \oplus TM$: If $(u_\alpha, M_\alpha)_{\alpha \in A}$ is an atlas of M, we may define an atlas $(Tu_\alpha \oplus Tu_\alpha, TM_\alpha \oplus TM_\alpha)_{\alpha \in A}$ for $TM \oplus TM$ by

$$Tu_\alpha \oplus Tu_\alpha: TM_\alpha \oplus TM_\alpha \to TU_\alpha \oplus TU_\alpha = U_\alpha \oplus \mathbb{R}^2 \oplus \mathbb{R}^2$$

$$(X, Y) \in T_pM \oplus T_pM \mapsto (u_\alpha(p), X_\alpha, Y_\alpha) \in U_\alpha \oplus T_{u_\alpha(p)}\mathbb{R}^2 \oplus T_{u_\alpha(p)}\mathbb{R}^2.$$

Here X_α and Y_α are representatives of X and Y with respect to the chart (u_α, M_α).

PROOF. Exactly analogous to the proof of (5.5.5). As with TM, the differentiable structure of $TM \oplus TM$ is completely determined by the differentiable structure of M. □

5.6.4 Definition. On a differentiable manifold,
 i) a 1-*form on M* is a differentiable mapping

$$\omega: TM \to \mathbb{R}$$

which has the following property: for every $p \in M$, $\omega|T_pM: T_pM \to \mathbb{R}$ is a 1-form, i.e., $\omega|T_pM \in T_p^*M := (T_pM)^*$;
 ii) a 2-*form on M* is a differentiable mapping

$$\Omega: TM \oplus TM \to \mathbb{R}$$

with the following property: for every $p \in M$, $\Omega|T_pM \oplus T_pM \in \Lambda^2 T_p^*M$.

Remarks. 1. What happens in local coordinates, i.e., terms of a chart (u_α, M_α) on M? A 1-form ω on M determines a 1-form $\omega_\alpha: TU_\alpha = U_\alpha \times \mathbb{R}^2 \to \mathbb{R}$ via $\omega_\alpha(X_\alpha) = \omega(X)$. Here X_α is the representative of X.
2. On U_α we have the natural coordinate basis $(e_1(u_\alpha), e_2(u_\alpha))$. Let $(du_\alpha^1, du_\alpha^2)$ denote the dual basis. The 1-form ω_α may be written as $\omega_\alpha = \sum_i a_i(u_\alpha) du_\alpha^i$, where $a_i(u_\alpha)$ are differentiable functions of u_α. Conversely ω_α determines a 1-form on M_α.
3. If (u_α, M_α) and (u_β, M_β) are two charts on M, then

$$d(u_\beta \circ u_\alpha^{-1}): T_{u_\alpha(p)}\mathbb{R}^2 \to T_{u_\beta(p)}\mathbb{R}^2: e_k(u_\alpha) \mapsto \sum_i \frac{\partial u_\beta^i}{\partial u_\alpha^k} e_i(u_\beta)$$

is a bijective linear map. By (5.6.1, 2), the dual mapping $d(u_\beta \circ u_\alpha^{-1})^*$ is given by

$$du_\beta^j \mapsto \sum_i \frac{\partial u_\beta^j}{\partial u_\alpha^i} du_\alpha^i.$$

5 Two-Dimensional Riemannian Geometry

Therefore, if $\omega_\beta = \sum_j b_j(u_\beta)\,du_\beta^j$ and $\omega_\alpha = \sum_l a_l(u_\alpha)\,du_\alpha^l$ are representatives of a 1-form ω with respect to two different charts, then

$$d(u_\beta \circ u_\alpha^{-1})^* \omega_\beta = \sum_{j,l} b_j(u_\beta) \frac{\partial u_\beta^j}{\partial u_\alpha^l}\,du_\alpha^l = \omega_\alpha,$$

$$a_l(u_\alpha) = \sum_j b_j(u_\beta) \frac{\partial u_\beta^j}{\partial u_\alpha^l}.$$

This last formula tells us how the components of 1-forms transform under a change of coordinates.

4. A 2-form Ω on a surface M is represented with respect to every chart (u_α, M_α) by $\Omega_\alpha\colon TU_\alpha \oplus TU_\alpha \to \mathbb{R}$, where $\Omega_\alpha(X_\alpha, Y_\alpha) = \Omega(X, Y)$, X_α, Y_α being the representatives of X and Y.

5. At each point of U_α the 2-form $du_\alpha^1 \wedge du_\alpha^2$ is a basis for the one-dimensional vector space of 2-forms: If $X_\alpha = \sum_i \xi_\alpha^i e_i$ and $Y = \sum_j \eta_\alpha^j e_j$, then

$$du_\alpha^1 \wedge du_\alpha^2 \colon (X_\alpha, Y_\alpha) \mapsto \xi_\alpha^1 \eta_\alpha^2 - \eta_\alpha^1 \xi_\alpha^2.$$

Therefore Ω_α may be written as $\Omega_\alpha = A(u_\alpha)\,du_\alpha^1 \wedge du_\alpha^2$, where $A(u_\alpha)$ is a differentiable function from U_α to $\mathbb{R}$.

6. How does $A(u_\alpha)$ transform under a change of coordinates? Suppose Ω is a 2-form whose representation with respect to (u_α, M_α) is $A(u_\alpha)\,du_\alpha^1 \wedge du_\alpha^2$ and with respect to (u_β, M_β) is $B(u_\beta)\,du_\beta^1 \wedge du_\beta^2$. By (5.6.1, 5),

$$\Lambda^2\,d(u_\beta \circ u_\alpha^{-1})^*\,du_\beta^1 \wedge du_\beta^2 = \det(d(u_\beta \circ u_\alpha^{-1}))\,du_\alpha^1 \wedge du_\alpha^2.$$

Therefore

$$\Lambda^2\,d(u_\beta \circ u_\alpha^{-1})^* \Omega_\beta = \Omega_\alpha; \qquad B(u_\beta)\det(d(u_\beta \circ u_\alpha^{-1})) = A(u_\alpha).$$

5.6.5 Proposition. (Definition of differential forms via local coordinates). *Differential forms may be defined in terms of local coordinates. Let $(u_\alpha, M_\alpha)_{\alpha \in A}$ be an atlas for M. Suppose for every $\alpha \in A$ a 1-form $\omega_\alpha \colon TU_\alpha \to \mathbb{R}$ is specified and that, for every $(\alpha, \beta) \in A \times A$,*

(*) $$d(u_\beta \circ u_\alpha^{-1})^* \omega_\beta = \omega_\alpha.$$

Then the 1-form $\omega \colon TM \to \mathbb{R}$ given by $\omega(X) = \omega_\alpha(X_\alpha)$, where X_α is the representation of X in the α-coordinate system, is well defined.

Similarly, given a 2-form $\Omega_\alpha \colon TU_\alpha \oplus TU_\alpha \to \mathbb{R}$ corresponding to each (u_α, M_α) such that

(**) $$\Lambda^2\,d(u_\beta \circ u_\alpha^{-1})^* \Omega_\beta = \Omega_\alpha$$

for all $(\alpha, \beta) \in A \times A$, the 2-form $\Omega \colon TM \oplus TM \to \mathbb{R}$ defined by $\Omega(X, Y) := \Omega_\alpha(X_\alpha, Y_\alpha)$, where X_α, Y_α are representatives of X and Y in the α-coordinate system, is well defined. The forms ω and Ω are differentiable since each ω_α and Ω_α is differentiable.

PROOF. The only possible problem arises when $Y, X \in TM$ are represented in two coordinate systems by X_α, Y_α and X_β, Y_β, respectively. Then "well defined" means $\omega_\alpha(X_\alpha) = \omega_\beta(X_\beta)$ and $\Omega_\alpha(X_\alpha, Y_\alpha) = \Omega_\beta(X_\beta, Y_\beta)$. But this follows immediately from the transformation "laws" (*) and (**) combined with the results of the previous section, (5.6.4). □

This method of defining forms is useful for constructing examples of differential forms.

5.6.6 Examples

1. Suppose $f: M \to \mathbb{R}$ is a differentiable function. The *differential of f* is the 1-form

$$df: TM \to \mathbb{R}, \qquad X \mapsto d(f \circ u_\alpha^{-1})X_\alpha,$$

where X_α = representative of X in TU_α. For example, the dual 1-forms du_α^1, du_α^2 corresponding to the basis vector fields $e_1(u)$, $e_2(u)$ of U_α are in fact the differentials of the coordinate functions $u_\alpha^k: M_\alpha \to \mathbb{R}$.

2. Suppose M has a Riemannian metric, g, given in each coordinate system (u_α, M_α) by (U_α, g_α). Using (5.6.1, 6), $T_p M$ is isomorphic to $T_p^* M = (T_p M)^*$ by means of the mapping

$$Lg_p: X \in T_p M \mapsto g_p(X, \;) \in T_p^* M.$$

Thus a vector field defines a 1-form and conversely: if X is represented locally by $X_\alpha = \sum_i \xi_\alpha^i(u_\alpha) e_i(u_\alpha)$, the 1-form corresponding to X is

$$\omega_\alpha = \sum_{i,j} g_{ij}(u_\alpha) \xi_\alpha^i(u_\alpha) \, du_\alpha^j.$$

3. Suppose M is oriented. Then we may define a 2-form, dM, called the *area-element* of M as follows: With respect to a positively oriented atlas $(u_\alpha, M_\alpha)_{\alpha \in A}$, $dM(u_\alpha) := \sqrt{g_\alpha(u_\alpha)} \cdot du_\alpha^1 \wedge du_\alpha^2$, where $g_\alpha(u_\alpha) := \det(g_{\alpha ij}(u_\alpha))$. Since $g_\beta(u_\beta)(\det(d(u_\beta \circ u_\alpha^{-1})))^2 = g_\alpha(u_\alpha)$, the transformation law (**) of (5.6.5) is satisfied and dM is well defined.

4. Suppose X is a vector field on M, M oriented. Then with respect to a positively oriented atlas $(u_\alpha, M_\alpha)_{\alpha \in A}$, the 1-form $i_X \, dM$ defined locally by

$$\omega_\alpha := \sqrt{g_\alpha}(-\xi_\alpha^2 \, du_\alpha^1 + \xi_\alpha^1 \, du_\alpha^2),$$

where $X_\alpha = \sum_j \xi_\alpha^j e_j(u_\alpha)$, is well defined. To see this, note that

$$d(u_\beta \circ u_\alpha^{-1})^* \, du_\beta^j = \sum_i \left(\frac{\partial u_\beta^j}{\partial u_\alpha^i}\right) du_\alpha^i,$$

$$\xi_\beta^i = \sum_k \frac{\partial u_\beta^i}{\partial u_\alpha^k} \xi_\alpha^k, \qquad \sqrt{g_\beta} = \sqrt{g_\alpha} \Big/ \det\left(\frac{\partial u_\beta^i}{\partial u_\alpha^k}\right).$$

Therefore $d(u_\beta \circ u_\alpha^{-1})^* \omega_\beta = \omega_\alpha$.

Remark. Sometimes $i_X \, dM$ is called the *interior product* of X with dM. One can easily see that $i_X \, dM$ coincides with the 1-form $Y \mapsto dM(X, Y)$. Moreover,

5 Two-Dimensional Riemannian Geometry

if X, Y form a positively oriented basis for T_pM,

$$i_X\, dM(Y) = dM(X, Y) = \sqrt{g(X, X)g(Y, Y) - g(X, Y)^2}.$$

One verifies this first for a local representation dM_α of dM and $(X, Y) = (e_1(u_\alpha), e_2(u_\alpha))$. The 1-form $i_X\, dM$ has a geometric interpretation. Given $X \in T_pM$, we define X' by

$$X' = Lg^{-1}i_X\, dM, \quad \text{i.e., } g(X', \ ) = i_X\, dM = dM(X, \ )$$

and claim that if $X \neq 0$, $\{X, X'\}$ is an orthogonal, positively oriented basis of T_pM and X' has the same length as X. Indeed, $g(X, X') = L_g X'(X) = dM(X, X) = 0$. Since $X' \neq 0$, either $dM(X, X')$ or $dM(X', X)$ is equal to $\sqrt{g(X, X)g(X', X')}$, depending on whether $\{X, X'\}$ or $\{X', X\}$ is positively oriented. But we know that $g(X', X') = dM(X, X') > 0$, hence $\{X, X'\}$ is positively oriented and $g(X', X') = \sqrt{g(X, X)g(X', X')}$.

We shall use this remark to obtain a geometric interpretation of Gauss' theorem (5.6.9).

The main reason for developing differential forms is that they allow us to define line and area integrals on a surface, M. We start by defining the two-dimensional analog of a piecewise smooth curve.

5.6.7 Definition. Let F denote a closed subset of the plane which is homeomorphic to the disk and whose boundary ∂F is a piecewise smooth simple closed curve whose exterior angles are all strictly less than π (see (2.1.5) for definition of exterior angle). A *(singular) polygon* on M is a smooth map $P: F \to M$. (If ∂F consists of three smooth curves, P is called a *(singular) simplex*.) The mapping $P|\partial F$ will be denoted by ∂P. If we consider ∂F a parameterized curve in the plane, ∂P parameterizes a piecewise smooth curve in M. We make the convention that ∂F will always be parameterized so that its rotation number is equal to $+1$. The coordinates on $F \subset \mathbb{R}^2$ will be denoted by (t^1, t^2).

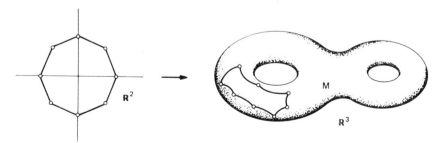

Figure 5.8 An example of a polygon on a surface M

5.6 Differential Forms

5.6.8 Definition (Integral of a 1-form on a curve). Given a 1-form ω on M and a curve $c: I \to M$, I compact, the integral $\int_c \omega$ is defined to be

$$\int_c \omega := \int_I \omega(\dot{c})\, dt.$$

(Integral of a 2-form on a polygon). Given a 2-form on M and a polygon $P: F \to M$, the integral $\iint_P \Omega$ is defined to be

$$\iint_P \Omega := \iint_F \Omega\left(\frac{\partial P}{\partial t^1}, \frac{\partial P}{\partial t^2}\right) dt^1\, dt^2.$$

Note that these integrals are invariant under an orientation-preserving change of variables of the curve or of the polygon.

We now state and prove the most important result of this section; Gauss' theorem relating the integral of the divergence of a vector field X on a polygon P to the line integral of $i_X\, dM$ on P. The next theorem generalizes the well-known Stokes theorem of two-dimensional calculus. When $M = \mathbb{R}^2$ with the standard metric, then $dM = du^1 \wedge du^2$, and the two theorems coincide.

5.6.9 Theorem (Gauss). *Let M be an oriented surface with a Riemannian metric. Let X be a vector field on M. Then for every polygon $P: F \to M$,*

$$\iint_P (\operatorname{div} X)\, dM = \int_{\partial P} i_X\, dM.$$

(*Note:* The last remark of (5.6.8) tells us that the integral on the right-hand side is well defined.)

PROOF.
1. Without loss of generality, we may assume that P lies entirely within one coordinate system. For if not, then we may subdivide F into $\{F_\rho\}$, $1 \le \rho \le k$, so that each $P_\rho := P|_{F_\rho}$ lies inside a coordinate system. Then if the theorem is true for each P_ρ,

$$\sum_\rho \iint_{P_\rho} \operatorname{div} X\, dM = \sum_\rho \int_{\partial P_\rho} i_X\, dM.$$

The left-hand side is equal to $\iint_P (\operatorname{div} X)\, dM$ by definition. The right-hand side is equal to $\int_{\partial P} i_X\, dM$ for the following reason: Every inner edge of $\bigcup_{1 \le \rho \le k} P_\rho$ appears twice and is traversed once in each direction. Therefore the integrals cancel out on each inner edge, leaving $(\operatorname{div} X)\, dM$ integrated over ∂P.

2. Suppose now that $P(F)$ lies entirely in one coordinate patch $u: U \to M$, $U \subset \mathbb{R}^2$. Then X may be written as $\xi^1 e_1 + \xi^2 e_2$. Therefore $\operatorname{div} X =$

117

$1/\sqrt{g}\, \Sigma_i\, (\partial/\partial u^i)(\sqrt{g}\, \xi^i)$, $dM = \sqrt{g}\, du^1 \wedge du^2$, and $i_X\, dM = -\xi^2 \sqrt{g}\, du^1 + \xi^1 \sqrt{g}\, du^2$. By the well-known Stokes theorem for two dimensions,

$$\iint_F \left(\frac{\partial f_2}{\partial u^1} - \frac{\partial f_1}{\partial u^2}\right) du^1\, du^2 = \int_{\partial F} f_1\, du^1 + f_2\, du^2.$$

When $f_1 = -\xi^2 \sqrt{g}$ and $f_2 = +\xi^1 \sqrt{g}$, this gives the required result. □

Remark. We can make a "physical" interpretation of the line integral in Gauss' theorem. Let ∂P be parameterized (locally) by $c(t)$, $|\dot{c}(t)| = 1$. Let $\{e_1(t) = \dot{c}(t), e_2(t)\}$ be the Frenet frame of $c(t)$. We may write $X|c(t)$ as $X(t) = \xi^1(t)e_1(t) + \xi^2(t)e_2(t)$. Then using the remark in (5.5.6, 4),

$$i_X\, dM(\dot{c}(t)) = g_{c(t)}(X'(t), \dot{c}(t)) = -\xi^2(t),$$

where $X'(t) = -\xi^2(t)e_1(t) + \xi^1(t)e_2(t)$. Therefore the integrand is equal to $-X(t) \cdot e_2(t)$. Since $e_2(t)$ is the inward pointing normal to P at $c(t)$, the line integral measures the "flow" of X out of the region P.

5.6.10 Suppose M is a compact oriented surface.
i) A *polygonal decomposition* Π of M is a finite family $\{P_\rho: F_\rho \to M\}$, $1 \le \rho \le k$, of orientation-preserving polygons on M (i.e., polygons on M such that P_ρ is an orientation-preserving differentiable mapping) which satisfy

$$\bigcup_\rho P_\rho(F_\rho) = M,$$

and if $r \ne s$, either $P_r \cap P_s$ is empty or consists of a corner of both polygons or an entire edge of both polygons. *Note:* A corner of $P_\rho(F_\rho)$ in this context is the image under P_ρ of a corner of the boundary of F_ρ.
ii) Given a 2-form, Ω, on M and a polygonal decomposition Π of M, we define

$$\iint_M \Omega := \sum_\rho \iint_{P_\rho} \Omega.$$

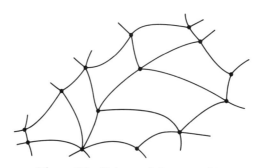

Figure 5.9 Polygonal decomposition

5.6.11 Proposition. *Suppose* $\Pi = \{P_\rho\}$ *and* $\Pi' = \{P'_{\rho'}\}$ *are two polygonal decompositions of a surface M as defined in (5.6.10, i). If Ω is a 2-form, then*

(*) $$\sum_\rho \iint_{P_\rho} \Omega = \sum_{\rho'} \iint_{P'_{\rho'}} \Omega.$$

This means that $\iint_M \Omega$, which is defined in terms of a polygonal decomposition, is in fact independent of the choice of such a decomposition.

PROOF. To prove (*), consider $\{P_\rho \cap P'_{\rho'}\}$ which is also a polygonal decomposition. Now each side of (*) is equal to $\sum_{\rho,\rho'} \iint_{P_\rho \cap P'_{\rho'}} \Omega$, and hence they are equal to each other. □

We conclude the chapter with a result that will have important applications in Chapter 6.

5.6.12 Theorem (Gauss' theorem for compact surfaces). *Let M be a compact surface with Riemannian metric. Then if X is a vector field on M,*

$$\iint_M (\text{div } X) \, dM = 0.$$

PROOF. Write the integral, in terms of some polygonal decomposition Π, as a sum and apply (5.6.9) to each summand. Each edge appears as a curve on two polygons, with opposite orientations, and the corresponding line integrals cancel each other.

5.7 Exercises and Some Further Results

5.7.1 The gradient. The *gradient* vector field of a differentiable function $\psi: M \to \mathbb{R}$ is the vector field $p \mapsto Lg_p^{-1} \, d\psi_p$. Recall that $Lg_p: T_pM \to T_p^*M$ is the isomorphism $X \mapsto g_p(X, \ )$ (see 5.6.1, 6). The gradient is usually denoted by grad $\psi(p)$ or simply grad ψ.

The gradient vector field generalizes the Euclidean notion of a gradient. If $U \subset \mathbb{R}^2$ is a subset of the plane and $\psi: U \to \mathbb{R}$ is a differentiable function, grad $\psi = (\partial \psi/\partial u^1)e_1 + (\partial \psi/\partial u^2)e_2$. If $X = x^1 e_1 + x^2 e_2$ is a tangent vector at $u_0 \in U$, the directional derivative of ψ in the direction X is equal to $X \cdot \text{grad } \psi(p) = x^1(\partial \psi/\partial u^1) + x^2(\partial \psi/\partial u^2)$. Similarly, if $\psi: M \to \mathbb{R}$ and $X \in T_pM$, then

$$d\psi(X) = g_p(X, \text{grad } \psi).$$

i) Suppose M is orientable and X is a vector field on M. Using (5.6.6), we may define dM and $i_X \, dM$. Prove: $i_{\psi X} \, dM = \psi i_X \, dM$ for any real-valued function $\psi: M \to \mathbb{R}$.

ii) In terms of a coordinate system (u_α, M_α) on M, show:

$$(\text{grad } \psi)_\alpha = \sum_{i,j} g^{ij}(u_\alpha) \frac{\partial \psi \circ u_\alpha^{-1}}{\partial u_\alpha^i} e_j(u_\alpha).$$

5 Two-Dimensional Riemannian Geometry

5.7.2 Let $\mathscr{F}(M)$ denote the set of differentiable functions on M. $\mathscr{F}(M)$ has a natural structure of an algebra over $\mathbb{R}$, the operations being defined pointwise. The *Laplace–Beltrami operator* is the mapping $\Delta \colon \mathscr{F}(M) \to \mathscr{F}(M)$ defined by

$$\Delta \psi := \operatorname{div} \operatorname{grad} \psi.$$

It is easy to convince yourself that if ψ is differentiable, then so is $\Delta \psi$, justifying the claim that $\Delta(\mathscr{F}(M)) \subset \mathscr{F}(M)$.

Prove:

$$\operatorname{div} \psi X = \psi \operatorname{div} X + d\psi X$$

$$\operatorname{div}(\psi_1 \operatorname{grad} \psi_2) = \psi_1 \Delta \psi_2 + g(\operatorname{grad} \psi_1, \operatorname{grad} \psi_2).$$

Use Gauss's theorem (5.6.9) to show

$$\iint_P \psi_1 \Delta \psi_2 + g(\operatorname{grad} \psi_1, \operatorname{grad} \psi_2) \, dM = \int_{\partial P} \psi_1 \frac{\partial \psi_2}{\partial n} \, ds,$$

where $\partial \psi_2 / \partial n = d\psi_2(n)$, and n is the outward pointing normal to the boundary curve P (see (5.6.7)). The above equation immediately implies

$$\iint_P \psi_1 \Delta \psi_2 - \psi_2 \Delta \psi_1 \, dM = \int_{\partial P} \left(\psi_1 \frac{\partial \psi_2}{\partial n} - \psi_2 \frac{\partial \psi_1}{\partial n} \right) ds.^3$$

These are *Green's formulae*.

5.7.3 A function $\psi \colon M \to \mathbb{R}$ is said to be *harmonic* if $\Delta \psi = 0$. It can be shown that each $p \in M$ has a neighborhood $\mathscr{M}$ on which a harmonic function $\psi \colon \mathscr{M} \to \mathbb{R}$ exists which satisfies $d\psi \neq 0$. The *conjugate harmonic function* χ is the harmonic function defined (up to sign) by the relations $g(\operatorname{grad} \psi, \operatorname{grad} \chi) = 0$, $g(\operatorname{grad} \chi, \operatorname{grad} \chi) = g(\operatorname{grad} \psi, \operatorname{grad} \psi)$.

If (u_α, M_α) is a chart such that (u_α^1, u_α^2) are conjugate harmonic functions, then the $u_\alpha = (u_\alpha^1, u_\alpha^2)$ are called *isothermal* (or *conformal*) *coordinates*. It was first proved by Lichtenstein[4] that isothermal coordinates always exist.

The line element in isothermal coordinates looks like

$$ds^2 = \frac{(du^1)^2 + (du^2)^2}{a(u^1, u^2)}, \qquad a(u^1, u^2) \neq 0.$$

Conversely, if the line element ds^2 has the above form, then the coordinate functions u^1 and u^2 are harmonic. This is because the Laplace–Beltrami operator for such a line element may be written in the form

$$\Delta \psi = \frac{((\partial^2 \psi / (\partial u^1)^2) + (\partial^2 \psi / (\partial u^2)^2))}{a(u^1, u^2)}.$$

---

[3] For further details about the Laplacian on a Riemannian manifold, see Berger *et al.* [B3].

[4] Lichtenstein, L. Beweis des Satzes, dass jedes hinreichend kleine, im wesentlichen stetig gekrümmte, singularitätenfreie Flächenstuck auf einen Teil einer Ebene zusammenhängend und in den kleinsten Teilen ähnlich abgebildet werden kann. Abh. Kgl. Preuss. Akad. Wiss. Berlin, Phys.-Math. Klasse, 1911, Anhang, Abhandlung VI, 1–49.

5.7 Exercises and Some Further Results

Suppose $\phi: (v^1, v^2) \mapsto (u^1, u^2)$ is a change of variables between isothermal coordinate systems (*a conformal mapping*). Prove: The functions $u^1(v^1, v^2)$, $u^2(v^1, v^2)$ satisfy the *Cauchy–Riemann equations*:

$$\frac{\partial u^1}{\partial v^1} = \frac{\partial u^2}{\partial v^2}, \qquad \frac{\partial u^1}{\partial v^2} = -\frac{\partial u^2}{\partial v^1}.$$

It follows from elementary complex analysis that ϕ can be written as a holomorphic function from an open set in the complex plane $\mathbb{C} = u^1 + iu^2$ into the complex plane $\mathbb{C} = v^1 + iv^2$.

In other words, the existence of isothermal coordinates on a surface M with a Riemannian metric implies that M can be given a complex structure, making it a Riemann surface in the sense of complex function theory.

5.7.4 More about minimal surfaces. Suppose $f: U \to \mathbb{R}^3$ is a parameterized surface in the sense of (3.1.1). If x, y, and z are the coordinates in $\mathbb{R}^3$, we may consider $x \circ f(u)$, $y \circ f(u)$, and $z \circ f(u)$ as functions on the surface M, represented in the coordinate chart and metric $(U, g) = (U, I)$. Here $g_u = I_u$ is the first fundamental form.
Show:

$$\Delta f(u) := (\Delta(x \circ f(u)), \Delta(y \circ f(u)), \Delta(z \circ f(u))) = 2H(u)n(u),$$

where H is the mean curvature and n is the unit normal vector field on the surface f.

Remark. This formula can be found in the proof of (6.2.9).

It follows from this formula that minimal surfaces, surfaces f which satisfy $H = 0$, are characterized by the fact that their three coordinate functions are harmonic.

Problem: State and prove an analogous result for minimal surfaces $f: U \to \mathbb{R}^n$ in Euclidean n-space.

Suppose (v^1, v^2) are local isothermal parameters on a minimal surface M in $\mathbb{R}^3$. (We may choose v^1 to be equal to one of the coordinate functions, e.g., $x \circ f(u)$. Then v^2 will be the harmonic conjugate of v^1. See (5.7.3).)
Show: The $\mathbb{R}^3$-valued function $f(v^1, v^2)$ is representable as the real part of a holomorphic function $F(v^2 + iv^2): \mathbb{C} \to \mathbb{C}^3$. (*Hint:* Use (5.7.3).)

Also, show: The Cauchy–Riemann equation simply that $F' = f_{v^1} - f_{v^2} -$ if$_{v^2} \neq 0$, $F'^2 = 0$. In other words, F is a holomorphic curve in $\mathbb{C}^3$ with $F'^2 = 0$. Conversely, if F is a holomorphic curve in $\mathbb{C}^3$ with $F' \neq 0$, $F'^2 = 0$, then its real part determines a minimal surface in $\mathbb{R}^3$.

This intimate connection between minimal surfaces and complex function theory is the basis for a highly developed theory of minimal surfaces.[5]

5.7.5 A *Liouville line element* on a surface M is a line element of the form

$$ds^2 = (A - B)(A_1^2\, du^2 + B_1^2\, dv^2), \qquad (u, v) \in U \subset \mathbb{R}^2,$$

where A, A_1 depend on u only and B, B_1 depend on v only.[6]

---

[5] See Nitsche, J. C. C. On new results in the theory of minimal surfaces. *Bull. Amer. Math. Soc.* **71**, 195–270 (1965); or the references of footnote 10 of chapter 3.

[6] See Darboux [A7], Part III, Book VI, Chapter 1.

5 Two-Dimensional Riemannian Geometry

Prove: The surfaces of revolution of (3.3.7, 3) are presented in coordinates which have a Liouville line element.

Prove: The lines of curvature on a surface of second order (3.7.3) determine a Liouville line element. For example, the line element on the ellipsoid is $ds^2 = \text{grad } \psi(v) \, dv^2 + \text{grad } \psi(w) \, dw^2$ (cf. (3.9.5)).

The most important property of a surface (U, g) with a Liouville line element is the existence of a nontrivial function $\Phi: T^0U \to \mathbb{R}$ which is constant on any one-parameter family $\{\dot c(t)\}$ of tangent vectors to a geodesic $c(t)$. As a special case of this we obtain Clairaut's theorem (4.5.1). Here $T^0U = \{X \in TU \mid X \neq 0\}$. If $X \in T^0U$, let

$$\Phi(X) := A(u(\pi X)) \cos^2 \theta(X) + B(v(\pi X)) \sin^2 \theta(X),$$

where $\theta(X)$ is equal to the angle between X and the tangent to the v-parameter curve.

Prove (Liouville's theorem): A curve $c(t) = (u(t), v(t))$ with $\dot u(t) \neq 0$ is (after possibly a reparameterization) a geodesic if and only if $\Phi(\dot c(t)) = $ constant.

Outline of proof. For an appropriate choice of a constant c, the functions

$$u' := \int A_1 \sqrt{A - c} \, du + \int B_1 \sqrt{c - B} \, dv$$

$$v' := \int \frac{A_1}{\sqrt{A - c}} \, du - \int \frac{B_1}{\sqrt{c - B}} \, dv$$

define a new coordinate system in which

$$ds^2 = du'^2 + (A - c)(c - B) \, dv'^2.$$

It then follows from (4.3.6) that the curves $v' = $ constant are geodesics, i.e., $\dot v' = (A_1/\sqrt{A - c})\dot u - (B_1/\sqrt{c - B})\dot v = 0$ implies that $(u(t), v(t))$ is (after possibly a reparameterization) a geodesic. The function Φ assumes the value c on the tangent vectors $(B_1\sqrt{A - c}, A_1\sqrt{c - B})$ to this geodesic.

Remark. On a surface with a Liouville line element, there exists two non-degenerate differentiable functions on T^0U which are constant on families $\{\dot c(t)\}$ of tangent vectors to geodesics $c(t)$. Namely $\Phi(X)$ and $g(X, X)/2$, the energy. In general, only the latter function exists. Given a surface with Riemannian metric, it is usually not possible to introduce coordinates whose line element is a Liouville line element.

The Global Geometry of Surfaces 6

In this chapter, we will consider some problems in the global differential geometry of surfaces. A "global" problem can be described as one which in general cannot be stated locally in terms of one coordinate system on a surface with a Riemannian metric, but must necessarily involve the total behavior of the surface. Most often, this total behavior is related to the topology of the surface. For example, Theorem (6.3.5) equates the integral of the curvature function $K(p)$ over a compact surface M with a topological invariant of M (the Euler characteristic). Neither of these two quantities can be described completely in terms of a single coordinate system.

Some of the theorems concern surfaces in Euclidean 3-space. Others treat abstract surfaces which are not realized in 3-space and are concerned entirely with intrinsically definable quantities.

When it is possible to do so without additional work, we will state and prove theorems for Riemannian manifolds. Otherwise, we will stick to surfaces and indicate what the appropriate generalization to manifolds would be.

6.1 Surfaces in Euclidean Space

6.1.1 A subset M of $\mathbb{R}^3$ is an *embedded surface* or simply a *surface* if, in the induced topology on M, there exists a family $(f_\alpha, U_\alpha)_{\alpha \in A}$ of parameterized surfaces $f_\alpha: U_\alpha \to \mathbb{R}^3$ in the sense of (3.1.1) satisfying
 i) each $f_\alpha: U_\alpha \to M_\alpha$ is a homeomorphism of U_α onto an open subset M_α of M;
 ii) the sets M_α cover M, i.e., $\bigcup_\alpha M_\alpha = M$.
  The homeomorphisms $u_\alpha = f_\alpha^{-1}: M_\alpha \to U_\alpha$ will be referred to as *charts* or *coordinate systems*, and will be denoted by (u_α, M_α).

6 The Global Geometry of Surfaces

It is easy to verify that the family $(u_\alpha, M_\alpha)_{\alpha \in A}$ defines a topological atlas for M. The next lemma will show that this is a differentiable atlas and M is a differentiable surface in the sense of (5.5.1).

Remark. More generally, we may define *embedded m-dimensional submanifolds* of $\mathbb{R}^n$ ($n \geq m$) to be subsets of $\mathbb{R}^n$ with the induced topology such that there exists a family $(f_\alpha, U_\alpha)_{\alpha \in A}$ (where U_α is an open set in $\mathbb{R}^m$ and $f_\alpha: U_\alpha \to \mathbb{R}^n$ is a regular map) satisfying (i) and (ii) above. The next lemma has a straightforward generalization to submanifolds. For clarity, we restrict ourselves to the case when $m = 2$ and $n = 3$.

6.1.2 Lemma. *Let M be a surface in $\mathbb{R}^3$ and let $(u_\alpha, M_\alpha)_{\alpha \in A}$ be an atlas for M as defined in (6.1.1).*
   i) *In the induced topology, M is a topological surface and the atlas $(u_\alpha, M_\alpha)_{\alpha \in A}$ is a differentiable atlas for M. Any two such atlases are equivalent.*
   ii) *The tangent space T_pM at $p \in M$ is represented by $T_{u_\alpha(p)}f \subset T_p\mathbb{R}^3$ whenever $p \in M_\alpha$. In particular, if $p \in M_\alpha \cap M_\beta$, $T_{u_\alpha(p)}f = T_{u_\beta(p)}f$. Therefore the restriction of the Euclidean inner product of $\mathbb{R}^3$ to T_pM is well defined. This inner product defines a Riemannian metric on M: in local coordinates*

$$g_{u_\alpha(p)}(*, *) = (df_\alpha)_{u_\alpha(p)}(*) \cdot (df_\alpha)_{u_\alpha(p)}(*).$$

PROOF. 1. As a subset of $\mathbb{R}^3$, the induced topology on M must be Hausdorff and have a countable basis. Moreover, for each $p \in M$ there exists an $\alpha = \alpha(p)$ such that $u_\alpha: M_\alpha \to U_\alpha \subset \mathbb{R}^2$ is a homeomorphism. All that remains to be shown is that the homeomorphism

$$u_\beta \circ u_\alpha^{-1}: u_\alpha(M_\alpha \cap M_\beta) \to u_\beta(M_\beta \cap M_\alpha)$$

is a diffeomorphism.

Let $p \in M_\alpha \cap M_\beta$. Now $f_\alpha = u_\alpha^{-1}: U_\alpha \to M_\alpha \subset \mathbb{R}^3$ is a parameterized surface (and therefore a regular map of constant rank $= 2$). We may apply the basic result (0.5.2) which asserts the existence of a neighborhood W of p in $\mathbb{R}^3$ and a diffeomorphism $g_\alpha: W \to W_\alpha$, where W_α is a neighborhood of $(u_\alpha(p), 0) \in \mathbb{R}^2 \times \mathbb{R}$, such that g_α satisfies

$$g_\alpha \circ f_\alpha(u_\alpha^1, u_\alpha^2) = (u_\alpha^1, u_\alpha^2, 0).$$

Consequently, if $U_\alpha(p)$ is a sufficiently small neighborhood of $u_\alpha(p) \in U_\alpha$, then $u_\alpha^{-1} | U_\alpha(p) = g_\alpha^{-1} | U_\alpha(p)$.

Similarly, there exists a diffeomorphism g_β from W onto a neighborhood W_β of $(u_\beta(p), 0) \in \mathbb{R}^2 \times \mathbb{R}$ so that $g_\beta \circ f_\beta$ is, locally, a linear injection. Therefore $u_\beta \circ u_\alpha^{-1}$ is equal to $g_\beta \circ g_\alpha^{-1} | U_\alpha(p)$ on a sufficiently small neighborhood of $u_\alpha(p)$. Since $g_\beta \circ g_\alpha^{-1}$ is a diffeomorphism and $u_\beta \circ u_\alpha^{-1}$ is equal to the restriction of $g_\beta \circ g_\alpha^{-1}$ to a linear subspace, $u_\beta \circ u_\alpha^{-1}$ is itself differentiable. Therefore $u_\beta \circ u_\alpha^{-1}$ is a diffeomorphism.

6.1 Surfaces in Euclidean Space

2. Let $p \in M_\alpha \cap M_\beta$. Suppose $f_\alpha | u_\alpha(M_\alpha \cap M_\beta)$ and $f_\beta | u_\beta(M_\alpha \cap M_\beta)$ are two parameterized surfaces (in the sense of Chapter 3) which are related by the change of variables $\phi_{\alpha\beta} = u_\alpha \circ (u_\beta | M_\beta \cap M_\alpha)^{-1} = f_\alpha^{-1} \circ f_\beta | u_\beta(M_\beta \cap M_\alpha)$, i.e., $f_\beta = f_\alpha \circ \phi_{\alpha\beta}$. According to (3.2.5), $g_{u_\alpha(p)} = I_{u_\alpha(p)}$ and $g_{u_\beta(p)} = I_{u_\beta(p)}$ define the equivalent metrics, as one would expect. This shows that the restriction of the Euclidean inner product to M defines a Riemannian metric on M as defined in (5.5.8). □

Examples of surfaces in $\mathbb{R}^3$ may be found in Chapter 3. In (3.3.7), the sphere f restricted to $(u, v) \in \,]-\pi/2, \pi/2[\, \times \,]-\pi, \pi[$, and the torus g restricted to $(u, v) \in \,]-\pi, \pi[\, \times \,]-\pi, \pi[$, are both surfaces. We have not considered the entire sphere or the entire torus in these examples. This is because the surfaces in Chapter 3 had to be defined in terms of a single coordinate system.

One of the most important ways in which surfaces in $\mathbb{R}^3$ and submanifolds of $\mathbb{R}^n$ arise are as the level sets of differentiable functions on $\mathbb{R}^3$ or $\mathbb{R}^n$. The next theorem describes sufficient conditions for the level sets

$$\{x \in \mathbb{R}^n \mid \psi(x) = c\}$$

of a differentiable map $\psi \colon \mathbb{R}^n \to \mathbb{R}^k$ to be submanifolds of $\mathbb{R}^n$.

6.1.3 Theorem. *Suppose D is an open set in $\mathbb{R}^n$ and $\psi \colon D \to \mathbb{R}^k$ is a differentiable map, where $0 \le k \le n$. If $a \in \psi(D)$ is a regular value of ψ (i.e., for all $p \in \psi^{-1}(a)$, $d\psi_p \colon T_p\mathbb{R}^n \to T_a\mathbb{R}^k$ is onto, or equivalently $d\psi_p$ has rank $= k$), then $M = \psi^{-1}(a)$ is an $(n - k)$-dimensional submanifold of $\mathbb{R}^n$.*

Note: M is not necessarily connected. Consider $\psi(x, y) = x^2 - y^2$ and $a \ne 0$.

PROOF. We will consider the case $n = 3$, $k = 1$, the case of surfaces in $\mathbb{R}^3$. The general case is similar (see Edwards, *Advanced Calculus of Several Variables*, Academic Press, 1973, pp. 196–200). We will show that, given $p \in M$, there exists a parameterized surface $f \colon U \to \mathbb{R}^3$ in the sense of Chapter 3 such that $f(U) = M'$ is an open subset of M containing p, and $f \colon U \to M'$ is a homeomorphism.

We may assume, without loss of generality, that $a = 0$, since we may replace ψ by $\psi - a$ without affecting the regularity of the function. Since $d\psi_p \colon T_p\mathbb{R}^3 \to T_0\mathbb{R}$ is onto, the implicit function theorem (0.5.2) asserts the existence of open neighborhoods V and V' of $p \in \mathbb{R}^3$ and a diffeomorphism $h \colon V \to V'$ such that $h(p) = p$ and $\psi \circ h \colon V \to \mathbb{R}$ is a linear, onto mapping of the form

$$(x^1, x^2, x^3) \in V \mapsto x^3 \in \mathbb{R}, \qquad \psi \circ h(p) = 0.$$

Let $M' = V' \cap M$ and $h^{-1}M' = U' \subset \mathbb{R}^2 = \{(x^1, x^2, x^3) \in \mathbb{R}^3 \mid x^3 = 0\}$. Then $f = h | U' \colon U' \to M' \subset M \subset \mathbb{R}^3$ is the desired parameterized surface. □

6　The Global Geometry of Surfaces

6.1.4 Example. $S_r^2(x_0) = \{x \in \mathbb{R}^3 \mid |x - x_0| = r\}, r > 0$, the sphere of radius r centered at $x_0 \in \mathbb{R}^3$.

Define $\psi \colon \mathbb{R}^3 \to R$ by $x \mapsto |x - x_0|^2$. First, note that r^2 is a regular value of ψ. If $\psi(x) = r^2$, then $d\psi_x(x - x_0) = \sum_i [(\partial/\partial x^i)(\sum_j (x^j - x_0^j)^2)]_x (x^i - x_0^i) = 2\sum_i (x^i - x_0^i)^2 = 2r^2 > 0$. The induced Riemannian metric on $S_r^2(x_0)$ can be written explicitly in terms of the chart $f^\tau \mid \,]-\pi/2, \pi/2[\, \times \,]-\pi, \pi[$ which was introduced in (4.4.5, 2). Since $f^\tau \cdot f^\tau = r^2$, it follows that f^τ is a chart for $S_r^2(0)$. The surfaces $S_r^2 = S_r^2(0)$ and $S_r^2(x_0)$ differ by a translation in $\mathbb{R}^3$ which does not disturb the geometry of the surface. All the calculations of the metric g, of the Gauss' curvature, etc., carry over without change. In particular, $K = 1/r^2$.

We can now give a second proof of the fact that geodesics on the sphere consist of arcs of great circles. To do this we will use the characterization of geodesics as curves which locally measure length (see (5.3.4)). Suppose c is a nontrivial, i.e., nonconstant, geodesic on $M = S_r^2$. Choose p_0 and p_1 two different points on c which are not antipodal points, and such that the arc c' of c which connects p_0 to p_1 is the unique length minimizing geodesic from p_0 to p_1 (cf. (5.2.5)).

Let d be the uniquely determined arc of a great circle which connects p_0 to p_1 whose length is strictly less than πr. We will now prove that, after possibly a reparameterization, $c' = d$.

The reflection σ of S_r^2 through the plane determined by d is an isometry which fixes d. In fact the only fixed points of S_r^2 are the points on the great circle determined by d, which includes p_0 and p_1. The length-minimizing geodesic c' connecting p_0 to p_1 is mapped into a length-minimizing geodesic $\sigma c'$ connecting $\sigma p_0 = p_0$ to $\sigma p_1 = p_1$. By the uniqueness of minimizing geodesics between p_0 and p_1, $\sigma c' = c'$, and therefore c' lies on the great circle determined by d. Therefore $c' = d$ up to parameterization.

More generally, $S_r^{n-1}(x_0) = \{x \in \mathbb{R}^n \mid |x - x_0| = r\}$, $r > 0$, the hypersphere in $\mathbb{R}^n$ of radius r and centered at x_0 is an $(n-1)$-dimensional submanifold of $\mathbb{R}^n$. It may also be shown that geodesics on $S_r^{n-1}(x_0)$ consist of segments of great circles. The proof is similar.

6.1.5 More examples. 1. *The torus.* Let

$$\psi(x) = \psi(x^1, x^2, x^3) := (\sqrt{(x^1)^2 + (x^2)^2} - a)^2 + (x^3)^2.$$

If $0 < b < a$, b^2 is a regular value of ψ. For if $x \in M = \psi^{-1}(b)$, then $(x^1)^2 + (x^2)^2 > 0$ and $d\psi_x(y) = 2b^2$ for the following value of $y = (y^1, y^2, y^3)$:

$$y^i := \frac{x^i \sqrt{(x^1)^2 + (x^2)^2} - a}{\sqrt{(x^1)^2 + (x^2)^2}}, \quad i = 1, 2, \ y^3 := x^3.$$

The values of the map $g(u, v)$, defined in (3.3.7), lie in M, so this is the familiar torus in Euclidean space. By changing to cylindrical coordinates, it is easily seen that this is a torus, symmetric about the (r, θ)-plane, with radii a and b.

2. *A surface of second order* (see (3.7.3)). Let $0 < c < b < a$, $\rho \notin \{a, b, c\}$, $\rho < a$. Let (x, y, z) denote the coordinates in $\mathbb{R}^3$. Define

$$\psi(x, y, z) := \frac{x^2}{c - \rho} + \frac{y^2}{b - \rho} + \frac{z^2}{a - \rho} - 1.$$

If $(x, y, z) \in \psi^{-1}(0)$, $d\psi(x, y, z) = 2\psi(x, y, z) + 2 = 2$. Therefore 0 is a regular value of ψ.

3. *Matrix groups.* If we identify $M_n = $ space of all $n \times n$ matrices with $\mathbb{R}^{n^2}$, then various classical groups appear as submanifolds of Euclidean space. First we restrict our attention to $GL(n, \mathbb{R}) = \{A \mid \det A \neq 0\}$. Since det: $M_n \to \mathbb{R}$ is a continuous function, $GL(n, \mathbb{R})$ is an open set in $\mathbb{R}^{n^2}$. By exercise (6.8.12) det: $GL(n, \mathbb{R}) \to \mathbb{R}$ is a differentiable function all of whose values are regular values. If ${}^t A$ denotes the transpose of A, consider the map $S: GL(n, \mathbb{R}) \to GL(n, \mathbb{R})$ given by $S(A) = {}^t A \cdot A$. Actually $S(A)$ is a symmetric matrix, so we may consider S as a map from $\mathbb{R}^{n^2}$ to $\mathbb{R}^{n(n+1)/2}$. Let I denote the identity matrix. Then $O(n) = \{A \in M_n \mid S(A) = I\}$, the *orthogonal* group, is a sub-manifold of dimension $n^2 - (n(n+1)/2) = (n(n-1)/2)$ in M_n, since I is a regular value of S. (The proof is left as an exercise.) The group $SO(n) = \{A \in O(n) \mid \det A = 1\}$ is called the *special orthogonal* group. It corresponds to orientation-preserving rotations of Euclidean n-space.

All the above submanifolds of $\mathbb{R}^{n^2}$ which we have been calling groups are indeed groups under matrix multiplication. For example, if $\det A \neq 0$ and $\det B \neq 0$, then $\det A \cdot B = \det A \cdot \det B \neq 0$, and A^{-1} exists and has nonzero determinant. Thus $GL(n, \mathbb{R})$ is a group. Similarly, if $\det A = \det B = +1$, then $\det A \cdot B = +1$ and if $S(A) = S(B) = I$ then $S(A \cdot B) = {}^t(AB)(AB) = {}^t B {}^t A A B = {}^t B B = I$, so $O(n)$ and $SO(n)$ are also subgroups of $GL(n, \mathbb{R})$. A group G with the structure of a differentiable manifold in which the mapping $(g, h) \in G \times G \mapsto gh^{-1} \in G$ is differentiable is called a Lie group. We may check that the above groups are Lie groups. The multiplication on $GL(n, \mathbb{R})$ is given in each coordinate by polynomials. Hence $(g, h) \mapsto gh^{-1}$ is a C^∞ map from $GL(n, \mathbb{R}) \times GL(n, \mathbb{R})$ to $GL(n, \mathbb{R})$. We have just shown that $O(n)$ is a closed sub-manifold of $GL(n, \mathbb{R})$. Hence $O(n) \times O(n) \subset GL(n, \mathbb{R}) \times GL(n, \mathbb{R})$ is a closed submanifold and the inclusion map is continuous. Thus the restriction of the map $(g, h) \mapsto gh^{-1}$ to $O(n) \times O(n)$ is a C^∞ map. For more details and an introduction to Lie groups, see Warner [B19] or Spivak [A17]. Cf. also (5.1.5.).

4. *Graphs of differentiable functions.* Let $A \subset \mathbb{R}^m$ be an open set and $f: A \to \mathbb{R}^n$ a differentiable function. The set

$$\text{graph } f := \{(x, y) \in \mathbb{R}^{m+n} \mid x \in A, y = f(x)\}$$

is a submanifold of $\mathbb{R}^{m+n}$. Indeed, graph f is the image of the regular map $f_\#: A \to \mathbb{R}^{m+n}$ (rank m) given by $f_\#(x) = (x, f(x))$.

An application of the implicit function theorem (0.5.2) shows that every submanifold of $\mathbb{R}^n$ of dimension k is locally the graph of a differentiable function f from some open set $A \subset \mathbb{R}^k$ into $\mathbb{R}^{n-k}$.

6 The Global Geometry of Surfaces

6.1.6 Proposition. *A surface $M \subset \mathbb{R}^3$ is orientable if and only if there exists a continuous function $n: M \to S^2 = S_1^2(0)$, $p \mapsto n(p)$, such that $n(p)$ is a unit normal vector to M at p., i.e., $n(p) \perp T_p M$.*

PROOF. 1. Let $(u_\alpha, M_\alpha)_{\alpha \in A}$ be a positive atlas for M. For $p \in M_\alpha$ define $n_\alpha(p)$ to be the Gauss normal vector to the surface $f_\alpha = u_\alpha^{-1}: U_\alpha \to \mathbb{R}^3$. If $p \in M_\beta$ we have $f_\beta = f_\alpha \circ (u_\alpha \circ u_\beta^{-1})$ and $\det d(u_\alpha \circ u_\beta^{-1}) > 0$ hence $n_\beta(p) = n_\alpha(p)$. (See the proof of (3.3.6).) Consequently, $n(p)$ is well defined and obviously continuous.

2. Conversely, suppose $n: M \to S^2$ is a continuous unit normal vector field as in the statement of the proposition. Let $(u_\alpha, M_\alpha)_{\alpha \in A}$ be any atlas of M. We construct a positive atlas out of this atlas as follows. The chart (u_α, M_α) remains unchanged if the Gauss normal vector field associated with $f_\alpha = u_\alpha^{-1}: U_\alpha \to \mathbb{R}^3$ agrees with n on M_n. If the Gauss normal vector field associated with f_α is equal to $-n$, then replace (u_α, M_α) with the chart $(s \circ u_\alpha, M_\alpha)$, where $s: \mathbb{R}^2 \to \mathbb{R}^2$ is the (orientation-reversing) reflection $(u^1, u^2) \to (-u^1, u^2)$. The new atlas is clearly an orientable atlas. □

Remark. If M is the level set at a regular value of a differentiable real valued function $\psi: \mathbb{R}^n \to \mathbb{R}$, as in (6.1.3), then $n(p) = \operatorname{grad} \psi(p)/|\operatorname{grad} \psi(p)|$ defines a unit normal vector field on M. Consequently, every component of M is orientable.

We may now extend the uniqueness theorem (3.8.8) for parameterized surface patches to oriented surfaces $M \subset \mathbb{R}^3$.

6.1.7 Proposition. *Suppose M and M^* are oriented and connected surfaces in $\mathbb{R}^3$. Then there exists an isometry B of $\mathbb{R}^3$ such that $BM = M^*$ if and only if there exists a diffeomorphism $\phi: M \to M^*$ which preserves the first fundamental form and preserves the second fundamental form up to sign.*

PROOF. 1. Using (6.1.6), we know it is possible to choose positive atlases $(u_\alpha, M_\alpha)_{\alpha \in A}$ and $(u_\beta^*, M_\beta^*)_{\beta \in B}$ for M and M^*, respectively, such that the Gauss normal vector fields n and n^* on the parameterized surface patches $f_\alpha: U \to \mathbb{R}^3$ and $f_\beta^*: U_\beta^* \to \mathbb{R}^3$ define global continuous mappings $n: M \to S^2$ and $n^*: M^* \to S^2$. Suppose $\phi: M \to M^*$ is a diffeomorphism satisfying the hypotheses of the proposition. Without loss of generality, we may assume that ϕ preserves second fundamental forms, for if necessary we may change the orientation of M^* and thereby change the sign of each n^* associated to $(u_\beta^*, M_\beta^*)_{\beta \in B}$. Applying (3.8.8) to each $f_\alpha: U_\alpha \to \mathbb{R}^3$, $\alpha \in A$, we may assert the existence of an isometry $B_\alpha: \mathbb{R}^3 \to \mathbb{R}^3$ such that $B_\alpha | M_\alpha = \phi | M_\alpha$. Since $B_\alpha | M_\alpha \cap M_{\alpha'} = \phi | M_\alpha \cap M_{\alpha'} = B_{\alpha'} | M_\alpha \cap M_{\alpha'}$ and M is connected, it follows that $B_\alpha = B_{\alpha'}$ for all $\alpha, \alpha' \in A$. Thus the required isometry exists.

2. Conversely, suppose B is an isometry of $\mathbb{R}^3$ with $BM = M^*$. It follows that $\phi = B | M$ is an isometry, $\phi: M \to M^*$. Certainly ϕ is one-to-one and onto. Given $\alpha \in A$, $f_\alpha = u_\alpha^{-1}: U_\alpha \to M_\alpha \subset \mathbb{R}^3$ is a local surface patch on

M and $B \circ f_\alpha: U_\alpha \to BM_\alpha \subset \mathbb{R}^3$ is a surface patch on M^*. By (6.1.2), $u_\beta^* \circ B \circ (u_\alpha \mid M_\alpha \cap B^{-1} M_\beta^*)^{-1}$ is a diffeomorphism. If we choose the sign of n so that $dBn = n^*$, it follows from (3.2.5) and the proof of (3.3.6) that the first and second fundamental forms are invariant under $\phi = B|M$. □

We end this section with our first result in global differential geometry, a result which is not only interesting in and of itself, but also has a number of useful applications.

6.1.8 Theorem. *On a compact surface $M \subset \mathbb{R}^3$, there must exist a point $p \in M$ where $K(p) > 0$.*

PROOF. Consider the continuous function $p \to |p|^2$ on M. By compactness of M, there exists a $p_0 \in M$ where this function assumes its maximum. Let $f: U \to \mathbb{R}^3$ be a local representation of M with $f(0) = p_0$. Locally, $|p|^2 = |f(u)|^2 = |f(0)|^2 + 2df_0(u) \cdot f(0) + d^2 f_0(u, u) \cdot f(0) + df_0(u) \cdot df_0(u) +$ 3rd and higher order terms. Since $f(0) = p_0$ is the point where the maximum value is obtained, $f(0) \cdot df_0 = 0$. Therefore $f(0) = \alpha \cdot n(0) \neq 0$, and the quadratic terms in the expansion of $|p|^2 = |f(u)|^2$ may be written in the form

$$\alpha II_0(u, u) + I_0(u, u) \leq 0.$$

Since I_0 is positive definite and $\alpha \neq 0$, αII_0 must also be a definite quadratic form. Therefore $K(p) = \det II_0 / \det I_0 > 0$. □

A somewhat more geometric but equivalent proof of this theorem goes as follows. Since $M \subset \mathbb{R}^3$ is compact, it lies inside the region bounded by some sphere $S_r^2(0)$ of sufficiently large radius centered at the origin. Let r shrink until $S_r^2(0)$ has a first point (or points) of contact with M. Let p_0 be one of these points of first contact. By exercise (6.8.12), all normal curvatures of M at p must have the same sign and have absolute value equal to or greater than $1/r$. It follows that $K(p) \geq 1/r^2$.

6.2 Ovaloids

In this section we will investigate a very interesting and important class of surfaces called *convex* surfaces. These are compact surfaces with strictly positive Gauss curvature. In $\mathbb{R}^3$, they turn out to be precisely the boundaries of bounded convex sets (see (6.2.3)).[1]

6.2.1 Definition. A compact surface $M \subset \mathbb{R}^3$ which has strictly positive Gauss curvature is called an *ovaloid*. In German, ovaloids are known as "Eifläche," literally egg-surfaces, a name apparently due to Blaschke [A2] and one that is quite suggestive of their appearance.

---

[1] Our convex surfaces are referred to by many mathematicians as "strictly convex" surfaces. The class of "convex" surfaces in this terminology includes those surfaces with $K \geq 0$.

6 The Global Geometry of Surfaces

In Chapter 2, section 3, we showed that a simply closed curve in the plane was convex (in the sense that it lay on one side of its tangent line at each point) if and only if its curvature was nonnegative. This result can easily be sharpened to say that a simply closed plane curve lies strictly on one side of each of its tangent lines if and only if its curvature is strictly positive. We shall prove analogous results for ovaloids.

EXAMPLE. The ellipsoid (3.7.3) with $\rho < c$, e.g., $\rho = 0$. To show that $K > 0$ is equivalent to showing that the second fundamental form, II, is definite (see (3.6.3)). To prove this, write the equation for the ellipsoid M in the form

$$\sum_{i=1}^{3} a_i(x^i)^2 = 1, \qquad a_3 > a_2 > a_1 > 0.$$

Let $x_0 \in M$. For $x \in M$ near to x_0 we may express the coordinates of x as follows:

$$x^i = x_0^i + \eta^i + \tfrac{1}{2}Q^i(\eta, \eta) + \text{third and higher order terms}.$$

Here Q^i is quadratic in $\eta = (\eta^1, \eta^2, \eta^3)$, and $x + \eta$ lies in the tangent space to M at x_0 (i.e., $\sum a_i x_0^i \eta^i = 0$).

The quadratic terms satisfy

$$\sum_i a_i x_0^i Q^i(\eta, \eta) + \sum a_i(\eta^i)^2 = 0.$$

This is a consequence of substituting $x^i = x_0^i + \eta^i + \ldots$ into $\sum_i^3 a_i(x^i)^2 = 1$ and looking at the quadratic terms. This relation shows that the normal component $\sum_i a_i x_0^i (x^i - x_0^i) = \sum_i a_i x_0^i Q^i(\eta, \eta) \neq 0$ whenever $\eta \neq 0$, hence $K > 0$.

6.2.2 Theorem (Hadamard's characterization of ovaloids).[2] *Suppose $M \subset \mathbb{R}^3$ is an ovaloid. Then*

i) *M is orientable;*
ii) *given an orientation of M, the normal map $n: M \to S^2$ which it defines is a diffeomorphism;*
iii) *M is strictly convex: for every $p \in M$, M lies entirely on one side of the tangent space $T_p M$; here $T_p M$ is considered as a plane through p in $\mathbb{R}^3$.*

PROOF. i) Since $K(p) > 0$, the second-order osculating surface to M at p is an elliptic paraboloid (see (3.6.3)). We can choose $n(p)$ to be the unit normal pointing in the direction of the positive axis of this paraboloid. The vector field $n(p)$ is clearly continuous and, by (6.1.6), M is orientable.

ii) In terms of a local representation $f: U \to \mathbb{R}^3$ of M, $K \neq 0$ is equivalent to the condition that $-dn_u$ or the Weingarten map $-dn_u \circ df_u^{-1}$ is bijective (see (3.5.5)). Therefore $dn_p: T_p M \to T_{n(p)} S^2$ is a bijection and, by the inverse function theorem (0.5.1), $n: M \to S^2$ is a local diffeomorphism.

[2] Hadamard, J. Sur certaines propriétés des trajectoires en dynamique. *J. Math. Pures Appl.* (5) **3**, 331–387 (1897). For a modern version of Hadamard's theorem which includes convex hypersurfaces in $\mathbb{R}^{n+1}$ (submanifolds of dimension n whose second fundamental form is positive definite), see Hopf [A12] or Chern [A6].

6.2 Ovaloids

Now this means that $n(M) \subset S^2$ is open and compact. It is certainly nonempty. Therefore $n(M) = S^2$, i.e., n is onto.

We will now prove that n is injective. Choose $p_0 \in M$ and let $p'_0 = n(p_0) \in S^2$. Let U_0 be an open neighborhood of p_0 and U'_0 a neighborhood of p'_0 chosen so that $n: U_0 \to U'_0$ is a diffeomorphism. Let m denote the inverse of n, that is, $m = (n|U_0)^{-1}: U'_0 \to U_0$. We claim that m may be extended to a continuous function $m: S^2 \to M$ which satisfies $n \circ m = \mathrm{id}$. It follows from this that $m(S^2) \subset M$ is open, nonempty and compact, which implies that m is surjective. Thus, if $n(p) = n(q)$, there exists p' and q' in S^2 with $m(p') = p$, $m(q') = q$. Applying n to these two equations implies that $p' = q'$ and $p = q$. Hence n is one-to-one.

The proof of the existence of the continuous extension of m to all of S^2 uses the well-known idea of monodromy from complex analysis. Let $p'_1 \in S^2$. Join p'_0 to p'_1 by means of a curve $c' = c'(t)$, $0 \leq t \leq 1$, on S^2 which has no self-intersections (for example, by a length-minimizing arc of a great circle = a length-minimizing geodesic).

Since $n: M \to S^2$ is a *local* diffeomorphism, it is possible to extend $m: U'_0 \to U_0$ to a mapping $m: U'(c') \to M$, where $U'(c')$ is a neighborhood of c', and $n \circ m = \mathrm{id}$. We prove this claim as follows. Suppose $t^* =$ the first value of t for which this is not possible. Certainly $t^* > 0$, since $c(0) = p'_0 \in U'_0$. Let $c(t) = m \circ c'(t)$, $0 \leq t < t^*$. As $t \to t^*$, $c(t)$ approaches a well-defined limiting value which we will denote by $c(t^*)$: for, since M is compact, there exist positive constants k, k' such that

$$k g_p(X, X) \leq g_{n(p)}(dn(X), dn(X)) \leq k' g_p(X, X)$$

for all $p \in M$ and all $X \in T_pM$. This means that a sequence $\{c'(t_i)\}$ on S^2, with $t_i < t^*$, $\lim_i t_i = t^*$, is Cauchy if and only if $\{c(t_i) = m \circ c'(t_i)\}$ is Cauchy in M.

For a suitably small neighborhood $U(c(t^*))$ of $c(t^*)$, $n|U(c(t^*))$ is a diffeomorphism of $U(c(t^*))$ with some subset of S^2 which contains $c'(t^*)$. Obviously this diffeomorphism extends past t^*. This means that $t^* = 1$ and we have the required extension of m to a neighborhood of c'.

Suppose $c'' = c''(t)$, $0 \leq t \leq 1$, is some other curve on S^2 connecting p'_0 to p'_1. There exists a homotopy c_s, $0 \leq s \leq 1$, with $c' = c_0$, $c'' = c_1$ (i.e., c_s is a continuously varying one-parameter family of curves which begins at c' and ends at c''). But since the value of $m(p'_1)$ is undisturbed by sufficiently small continuous changes of the curve by which it is defined (a small change of c' still remains in $U'(c^*)$), it follows that $m(p'_1)$ is defined independent of the choice of the curve c'. This proves the existence of a globally defined inverse of n, from which (ii) follows.

To prove (iii) consider the "support function" $h: M \to \mathbb{R}$ of M at any $p_0 \in M$ which is defined by $p \mapsto h(p) := n(p_0) \cdot (p - p_0)$. The statement of (iii) is equivalent to proving that h does not change sign. Since M is compact, h assumes a minimum at some $p_1 \in M$. At p_1, $0 = dh_{p_1} = n(p_0) \cdot dp_{p_1}$, i.e.,

$n(p_0) = \pm n(p_1)$. If $p_1 \neq p_0$, then, by (ii), $n(p_0) = -n(p_1)$. This means that $h(p) = -n(p_1) \cdot (p - p_0)$. We may write

$$h(p) = -n(p_1) \cdot (p - p_0) = -n(p_1) \cdot (p - p_1 + p_1 - p_0)$$
$$= -n(p_1) \cdot (p - p_1) + h(p_1).$$

For values of p near p_1, the first term is negative by our choice of n, contradicting the choice of p_1 as the point where h assumes its minimum. Therefore $p_1 = p_0$ and $h(p) \geq 0$ for all $p \in M$. Moreover $h(p) > 0$ if $p \neq p_0$. For if $h(p) = 0$ then h assumes its minimum value at p and the previous argument applies. □

6.2.3 Corollary. Given $p \in M$, a surface and $n(p)$ a unit normal to M at p, let $\mathcal{H}_p := \{q \in \mathbb{R}^3 \mid n(p) \cdot (q - p) \geq 0\}$. $\mathcal{H}_p$ is the closed half-space bounded by T_pM and containing the point $p + n(p)$. If M is an ovaloid, let $K = \bigcap_{p \in M} \mathcal{H}_p$. By the previous theorem, part (iii), $M \subset K$. K is a convex set. That is, if $r, s \in K$ the line segment, $\overline{rs}$, joining r to s is also in K. Furthermore, if $\mathring{K}$ denotes the interior of K, $\mathring{K}$ is not empty and $M = K - \mathring{K}$. If $r \in \mathring{K}$ then $n(p) \cdot (p - r) < 0$ for all $p \in M$.

PROOF. Since $\mathcal{H}_p$ is convex for each $p \in M$, so is $K = \bigcap_{p \in M} \mathcal{H}_p$. Further, any $p \in M$ cannot be in $\mathring{K}$ because each neighborhood (in $\mathbb{R}^3$) of p must contain points which do not belong to $\mathcal{H}_p$. Thus $\mathring{K} \subset K - M$.

Given $p_0 \in K$, we have, for all $p \in M$, $n(p) \cdot (p - p_0) \leq 0$. If there is a $p_1 \in M$ with $n(p_1) \cdot (p_1 - p_0) = 0$ then $p_1 = p_0$. Indeed, otherwise there exist $p_1' \in M$ near p_1 with $n(p_1') \cdot (p_1' - p_0) > 0$ since $d(n(p) \cdot (p - p_0))_{p_1} = dn_{p_1} \cdot (p_1 - p_0) + n(p_1) \cdot dp_{p_1} = dn_{p_1} \cdot (p_1 - p_0) \neq 0$.

To show that $K - M \neq \emptyset$ we prove: If $p_0' \in M$ and $\epsilon = \epsilon(p_0') > 0$ sufficiently small then, for $p_0 = p_0' + \epsilon n(p_0')$, $g(p) = n(p) \cdot (p - p_0) < 0$ for all $p \in M$. Let $p_1 \in M$ such that $\alpha = g(p_1) \geq g(p)$ for all $p \in M$. Assume $\alpha \geq 0$. Since $0 = dg_{p_1} = dn_{p_1} \cdot (p_1 - p_0)$, we have $p_1 - p_0 = \alpha n(p_1)$. From $g(p) = n(p) \cdot (p - p_0) = n(p) \cdot (p - p_0') + \epsilon n(p) \cdot n(p_0') \leq \alpha$ it follows that

$$|p_1 - p_0'| \leq |p_1 - p_0| + |p_0 - p_0'| \leq \alpha + \epsilon \leq 2\epsilon.$$

Hence, if $\epsilon \to 0$, $p_1 \to p_0'$. On the other hand, for p_1 near p_0',

$$g(p_1) = n(p_1) \cdot (p_1 - p_0) = n(p_1) \cdot (p_1 - p_0') - \epsilon n(p_1) \cdot n(p_0') < 0,$$

a contradiction to $g(p_1) \geq 0$.

Finally, to prove that $p_0 \in K - M$ implies $p_0 \in \mathring{K}$ we observe that $\sup_{p \in M} n(p) \cdot (p - p_0) < 0$. This is true also for p' sufficiently near p_0. □

6.2.4 Definition. Let (k_{ij}) be a symmetric 2×2 matrix. The *adjoint transposed matrix* of (k_{ij}) is the matrix

$$(\check{k}^{ij}) := \begin{pmatrix} k_{22} & -k_{12} \\ -k_{12} & k_{11} \end{pmatrix}.$$

Clearly $\sum_j k_{ij}\check{k}^{jl} = \delta_i^l \det(k_{ij})$. Thus, if $k = \det(k_{ij}) \neq 0$, then $(\check{k}^{ij}/k)$ is the inverse matrix of (k_{ij}).

In preparation for the next theorems of this section, we prove two algebraic lemmata.

6.2.5 Lemma. *If (k_{ik}) and $(\tilde{k}_{lm})$ are related by the equations $\tilde{k}_{lm} = \sum_{i,k} a_l^i a_m^k k_{ik}$, where $a := \det(a_k^i) \neq 0$, then their respective adjoint transposes are related by*

$$\check{k}^{ik} = \sum_{l,m} b_l^i b_m^k \check{\tilde{k}}^{lm} \cdot a^2,$$

where $\sum_j b_j^i a_k^j = \delta_k^i$.

PROOF. Compute. In the case where $k = \det(k_{ik}) \neq 0$, this follows from the usual transformation law relating the inverse matrices of (k_{ik}) and $(\tilde{k}_{lm})$ (see (5.6.1, 2)):

$$\frac{\check{k}^{ik}}{k} = \sum_{l,m} b_l^i b_m^k \frac{\check{\tilde{k}}^{lm}}{\tilde{k}},$$

where $\tilde{k} = a^2 k$. $\square$

6.2.6 Lemma. *Suppose (h_{ik}) and (h_{ik}^*) are 2×2 positive definite symmetric matrices with $\det(h_{ik}) = \det(h_{ik}^*)$. Then $\det(h_{ik} - h_{ik}^*) \leq 0$ with equality if and only if $(h_{ik}) = (h_{ik}^*)$.*

PROOF. Both the hypotheses and the conclusions of this lemma are independent of transformations of the form $\tilde{h}_{lm} = \sum_{i,k} a_l^i a_m^k h_{ik}$ (or $\tilde{H} = AH^tA$), where $\det A \neq 0$. It follows from our results (3.5.2) and (3.5.3) on the principal curvature directions that a matrix $A = (a_k^i)$ exists such that

$$\tilde{H} = AH^tA = \begin{pmatrix} 1 & 0 \\ 0 & 1 \end{pmatrix}; \qquad \tilde{H}^* = AH^{*t}A = \begin{pmatrix} a_1 & 0 \\ 0 & a_2 \end{pmatrix}.$$

To see this, we consider H as the fundamental matrix of a first fundamental form and H^* as the fundamental matrix of a second fundamental form. Then a pair $\{X_1, X_2\}$ of H-orthonormal vectors exists for which

$$X \mapsto H^*(X, X); \qquad H(X, X) = 1$$

assumes its maximum at minimum, respectively.

Take as A the matrix carrying the natural basis of $\mathbb{R}^2$ into $\{X_1, X_2\}$. Then $\det(H - H^*) = (1 - a_1)(1 - a_2) = -(\sqrt{a_1} - \sqrt{a_2})^2$, since $a_1 a_2 = \det \tilde{H}^* = \det \tilde{H} = 1$ and $a_1 > 0$, $a_2 > 0$. This proves the lemma. $\square$

Geometric application of the above results

1. Let M be an orientable surface in $\mathbb{R}^3$ and let (u, M') be a chart on M. With respect to $f = u^{-1}: U \to M' \subset \mathbb{R}^3$, define the coefficients $h_{ik}(u)$ of the

6 The Global Geometry of Surfaces

second fundamental form. If (v, M'') is another similarly oriented chart and $\tilde{h}_{lm}(v)$ are the corresponding coefficients of II, then for $u \in u(M' \cap M'')$,

$$\tilde{h}_{lm}(v(u)) = \sum_{i,k} \frac{\partial u^i}{\partial v^l} \frac{\partial u^k}{\partial v^m} h_{ik}(u).$$

If we write $A = (\partial u^i / \partial v^l)|_v$, this equation may be written as

$$\tilde{H} = AH^t A.$$

The determinants of the first fundamental forms are related by

$$\tilde{g}(v(u)) = (\det A)^2 \cdot g(u).$$

Lemma (6.2.5) yields

$$\sum_{i,k} \tilde{h}_{ik} \frac{\tilde{h}^{ik}}{\tilde{g}} \bigg|_{v(u)} = \sum_{i,k} h_{ik} \frac{h^{ik}}{g} \bigg|_u.$$

I.e., this expression is independent of change of coordinates. Thus it gives a globally well-defined function on M.

2. Let M and M^* be two orientable surfaces in $\mathbb{R}^3$. Suppose $\phi: M \to M^*$ is an isometry, i.e., a diffeomorphism which preserves the first fundamental forms. If (u, M') is a chart for M, then $(u \circ \phi^{-1}, \phi M')$ is a chart for M^*. So we may write M and M^* locally in terms of the same parameters. Since ϕ is an isometry, $g_{ik}(u) = g_{ik}^*(u)$. Let $h_{ik}^*(u(p)) = h_{ik}^*(u \circ \phi^{-1}(\phi(p)))$. This defines a function on M. By (1) above,

(*) $$k(p) := \sum_{i,k} h_{ik} \frac{h^{*ik} - h^{ik}}{g} \bigg|_{u(p)}$$

is a well-defined function on M. It represents twice the difference of the Gauss curvature functions. We shall need $k(p)$ for

6.2.7 Lemma (The Herglotz integral formula[3]). *Suppose M and M^* are ovaloids in $\mathbb{R}^3$. By (6.2.2), both M and M^* are orientable, which means that after a choice of orientation the mean curvature functions H and H^* are defined. Suppose there exists $\phi: M \to M^*$, an isometry. Fix $x_0 \in \mathbb{R}^3$. Then the following integral formula holds:*

(**) $$\iint_M k(p) n(p) \cdot (p - x_0) \, dM = \iint_M 2H(p) \, dM - \iint_M 2H^*(\phi(p)) \, dM,$$

where $k(p)$ is defined in (*) *above.*

PROOF. 1. Let (u, M') be a chart for M and $(u \circ \phi^{-1}, \phi M')$ the associated chart of M^*. If $f = u^{-1}: U \to \mathbb{R}^3$, we will show that

(1) $$\frac{1}{\sqrt{g}} \sum_{i,k} \frac{\partial}{\partial u^i} \left(\sqrt{g} \frac{h^{*ik}}{g} f_k \right) = \sum_{i,k} \frac{h^{*ik}}{g} h_{ik} n.$$

[3] Herglotz, G. Über die Starrheit der Einflächen. *Abh. Math. Sem. Univ. Hamburg* **15**, 127–129 (1943).

134

Both sides of (1) are invariant under change of coordinates: For the right-hand side this was shown above. If we write $f_k = \sum_r f_k^r e_r$ and let $\xi^i = \sum_k \check{h}^{*ik} f_k^r/g$, then the left-hand side equals div X, where $X = \sum_i \xi^i f_i$ (see (4.1.7)). Therefore we choose to verify (1) in Fermi coordinates. In such coordinates $\Gamma^l_{ik}(u_0) = 0$, $g(u_0) = 1$, and $\partial g/\partial u^i(u_0) = 0$, which makes the left-hand side of (1) equal to

(2) $$\sum_{i,k} \check{h}^{*ik}_{,i} f_k + \sum_{i,k} \check{h}^{*ik} h_{ik} n.$$

We apply the Mainardi–Codazzi equations (3.8.3 (ii)): $h_{ik,l} - h_{il,k} = 0$, which imply

$$\sum_i \check{h}^{*ik}_{,i} = \begin{cases} \check{h}^{*11}_{,1} + \check{h}^{*21}_{,2} = h^*_{22,1} - h^*_{12,2} = 0 & (k = 1) \\ \check{h}^{*12}_{,1} + \check{h}^{*22}_{,2} = -h^*_{12,1} + h^*_{11,2} = 0 & (k = 2). \end{cases}$$

This in turn implies that the first term of (2) is equal to zero. Therefore the left- and right-hand sides of (1) are equal.

2. Using (1),

(3)
$$\sum_{i,k} \frac{\check{h}^{*ik}}{g} h_{ik} n \cdot (f - x_0) = \frac{1}{\sqrt{g}} \left(\sum_{i,k} \frac{\partial}{\partial u^i} \left(\sqrt{g} \frac{\check{h}^{*ik}}{g} f_k \right) \right) \cdot (f - x_0)$$
$$= -\frac{1}{g} \sum_{i,k} \check{h}^{*ik} g_{ik} + \frac{1}{\sqrt{g}} \sum_{i,k} \frac{\partial}{\partial u^i} \left(\sqrt{g} \frac{\check{h}^{*ik}}{g} f_k \cdot (f - x_0) \right).$$

Since

$$\frac{1}{g} \sum_{i,k} \check{h}^{*ik} g_{ik} = \frac{1}{g} (h^*_{22} g_{11} - 2h^*_{12} g_{12} + h^*_{11} g_{22}) = \sum_{i,k} h^*_{ik} g^{*ik} = 2H^*$$

and the second sum on the right-hand side in (3) may be written as div X, where $X = \sum_i \xi^i f_i$ and $\xi^i = \sum_k (\check{h}^{*ik}/g) f_k \cdot (f - x_0)$, integrating (3) over M yields

(4) $$\iint_M \sum_{i,k} \frac{\check{h}^{*ik}}{g} h_{ik} n \cdot (f - x_0) \, dM = -\iint_M 2H^* \, dM.$$

($\int_M$ div $X = 0$ by Gauss's theorem (5.6.12).) If $M = M^*$ the formula is true with the * deleted. Call this formula (4'). The difference between this formula (4) and (4') is (**). □

We now use the Herglotz integral formula to establish a famous result due to Cohn-Vossen and Herglotz.[4]

---

[4] Cohn-Vossen proved this theorem for analytic surfaces in 1927. The methods he employed were different from those presented here. Herglotz' proof is in the paper in footnote 3. He makes use of an idea of Blaschke. See Blaschke, W. *Über eine geometrische Frage von Euklid bis heute*. Hamburger Mathematische Einzelschriften, 23. H. Leipzig and Berlin: Teubner, 1938. For a discussion of the history of this problem as well as some further results in this area of research, see Efimov, N. W. *Flächenverbiegung im Großen; mit einem Nachtrag von E. Rembs und K. P. Grotemeyer*. Berlin: Akademie-Verlag, 1957.

6 The Global Geometry of Surfaces

6.2.8 Theorem (Rigidity of ovaloids). *Let M and M^* be two ovaloids which are isometric, i.e., there exists an isometry $\phi: M \to M^*$. Then there exists an isometry B of Euclidean 3-space which maps M onto M^* and which satisfies $B|M = \phi$.*

PROOF.

1. Using ϕ, we may introduce coordinates simultaneously on M and M^*. Choose normal vector fields n on M and n^* on M^* so that II and II^* are both positive definite (see the proof of (6.2.2)). Since ϕ is an isometry, $K^* = K$, $g^* = g$, and therefore $h = \det(h_{ik}) = K \cdot g = K^* \cdot g^* = h^*$. We claim that $(h_{ik}) = (h_{ik}^*)$. The theorem follows from this claim by an application of (6.1.7). By (6.2.6), the *claim* will follow if we can show that $\det(h_{ik} - h_{ik}^*) = 0$.

2. By (6.2.3) there exists an $x_0 \in \mathbb{R}^3$ for which $n(p) \cdot (p - x_0) < 0$, for all $p \in M$. Also,

$$\sum_{i,k}(\tilde{h}^{*ik} - \tilde{h}^{ik})h_{ik} = -2\det(h_{ik}) + h_{22}^* h_{11} - 2h_{12}^* h_{12} + h_{11}^* h_{22}$$
$$= -\det(h_{ik}^* - h_{ik}) \geq 0$$

by (6.2.6). Therefore $k(p) \geq 0$ and (6.2.7) implies

$$\iint_M 2H\,dM - \iint_M 2H^*\,dM \leq 0.$$

Since we could interchange M and M^* and derive the analogous inequality, it must be that

$$\iint k(p)n(p) \cdot (p - x_0)\,dM = 0.$$

The fact that $n(p) \cdot (p - x_0) < 0$ implies that $k(p) = 0$, hence

$$\det(h_{ik} - h_{ik}^*) = 0. \qquad \square$$

6.2.9 Lemma (The Minkowski integral formulae).[5] *Let M be a compact orientable surface in $\mathbb{R}^3$. Then the following integral formulae hold:*

i) $$-\iint_M H(p)n(p) \cdot (p - x_0)\,dM = \iint_M dM.$$

ii) $$-\iint_M H(p)\,dM = \iint_M K(p)n(p) \cdot (p - x_0)\,dM.$$

PROOF. 1. Let $f: U \to \mathbb{R}^3$ be a local representation of M by a positively oriented chart. We show that

(a) $$\sum_{i,k} \frac{1}{\sqrt{g}} \frac{\partial}{\partial u^i}(\sqrt{g} g^{ik} f_k) = 2Hn.$$

Since both sides of (a) are clearly invariant under choice of coordinates, it suffices to prove (a) in Fermi coordinates where $g_{ik,l}(u_0) = 0$ and $g_{ik}(u_0) = \delta_{ik}$. The left-hand side is then equal to $\sum_{i,k} g^{ik} h_{ik} n = 2Hn$ by definition.

[5] Minkowski, H. Volumen und Oberflache. *Math. Ann.* **57**, 447–495 (1903).

2. Taking the inner product of (a) with $(f - x_0)$, we get

$$Hn\cdot(f - x_0) = \frac{1}{2\sqrt{g}} \sum_{i,k} \frac{\partial}{\partial u_i} \{(\sqrt{g} g^{ik} f_k)\cdot(f - x_0)\} - \frac{1}{2\sqrt{g}} \sum_{i,k} \sqrt{g} g^{ik} g_{ik},$$

The last term is equal to -1. Statement (i) follows from Gauss's theorem (5.6.9).

3. To prove (ii) we proceed in a similar fashion. First, if f is a local representation of M,

(b) $$\frac{1}{\sqrt{g}} \sum_{i,k} \frac{\partial}{\partial u^i}\left(\sqrt{g}\frac{\check{h}^{ik}}{g} f_k\right) = 2Kn.$$

To prove this, we use the fact (established in the proof of (6.2.7)) that $g_{,i} = \sum_i \check{h}^{ik}_{;i} = 0$. Therefore the left-hand side of (b) is equal to

$$\frac{1}{g}\sum_{i,k} \check{h}^{ik} h_{ik} n = 2\left(\frac{h}{g}\right) n = 2Kn.$$

4. Taking the inner product of (b) with $(f - x_0)$,

$$Kn\cdot(f - x_0) = \frac{1}{2\sqrt{g}} \sum_{i,k} \frac{\partial}{\partial u^i}\left(\sqrt{g}\frac{\check{h}^{ik}}{g} f_k \cdot (f - x_0)\right) - \frac{1}{2\sqrt{g}}\sqrt{g}\sum_{i,k}\frac{\check{h}^{ik}}{g} g_{ik}.$$

The last term is equal to $-H$. Statement (ii) now follows from Gauss's theorem (5.6.9). $\square$

We end this section with a famous result of Liebmann[6] which characterizes the sphere as the only compact connected surface in $\mathbb{R}^3$ with constant curvature.

6.2.10 Theorem. *Let M be a compact connected surface in $\mathbb{R}^3$ with $K = $ constant. Then $K = r^2 > 0$ and $M = S^2_{1/r}$, a sphere of radius $1/r$.*

PROOF. 1. By (6.1.8), M must have at least one point of positive curvature and therefore $K > 0$. Setting $K = r^2 = $ constant,

$$\frac{1}{g}\det(h_{ik} - rg_{ik}) = K - 2rH + r^2 = 2r^2 - 2rH.$$

$$\Delta_0(r) := \iint_M \frac{1}{g}\det(h_{ik} - rg_{ik})\, dM = 2r^2 \iint_M dM - 2r \iint_M H\, dM$$

$$\Delta_1(r) := \iint_M \frac{1}{g}\det(h_{ik} - rg_{ik}) n\cdot(p - x_0)\, dM$$

$$= 2\iint_M Kn\cdot(p - x_0)\, dM - 2r \iint_M Hn\cdot(p - x_0)\, dM$$

$$= -2\iint_M H\, dM + 2r \iint_M dM \quad \text{(by (6.2.9))}.$$

Thus $\Delta_0(r) = r\Delta_1(r)$.

---

[6] Liebmann, H. Eine neue Eigenschaft der Kugel, in *Nachr. Kgl. Ges. Wiss. Göttingen, Math.-Phys. Klasse*, 44–55 (1899). For further references, see the book of Efimov referred to in footnote 4.

6 The Global Geometry of Surfaces

2. Since $K > 0$ we may choose a normal vector field n making II positive definite. Since M is an ovaloid, there exists an $x_0 \in \mathbb{R}^3$ for which $n(p) \cdot (p - x_0) < 0$ for all $p \in M$ (see (6.2.3)). Since $\det(h_{ik}) = K \cdot \det(g_{ik}) = r^2 \cdot \det(g_{ik}) = \det(rg_{ik})$, (6.2.6) implies that $\det(h_{ik} - rg_{ik}) \leq 0$, with equality, if and only if $h_{ik} = rg_{ik}$. Therefore $\Delta_0(r) \leq 0$ and $\Delta_1(r) \geq 0$ which, combined with the equality $\Delta_0(r) = r\Delta_1(r)$, implies that $\Delta_0(r) = \Delta_1(r) = 0$. Therefore $h_{ik} = rg_{ik}$, where r is a positive constant. This means that M consists entirely of umbilic points and, by (3.5.11), M must be a sphere of radius $= 1/r$. □

6.3 The Gauss–Bonnet Theorem

In this section we will prove one of the most important results in the global theory of surfaces. In contrast to the results in (6.2), which deal with the surfaces in Euclidean space, the Gauss–Bonnet theorem is a theorem of intrinsic differential geometry. In order to appreciate its full significance, some familiarity with the topology of compact orientable surfaces is necessary. This may be found in Seifert and Threlfall, *Lehrbuch der Topologie*, Chelsea, New York, N.Y., Lefschetz, S., *Introduction to Topology*, Princeton University Press, Princeton, N.J., 1949, or Massey, W. S., *Algebraic Topology*, Harcourt Inc, New York, N.Y., 1967.

In preparation for the proof of Theorem (6.3.2), consider a coordinate system (U, g) with $ds^2 = (du^1)^2 + g_{22}(du^2)^2$, i.e., geodesic coordinates. In this situation, $E_1(u) = (\partial/\partial u^1)(u)$, $E_2(u) = (\partial/\partial u^2)/\sqrt{g_{22}}$ is an orthonormal 2-frame on (U, g). Let $u(t)$, $t \in I$, be a curve in (U, g) with $g(\dot{u}, \dot{u}) = 1$. If $e_1(t), e_2(t)$ is the Frenet frame on $u(t)$, then the geodesic curvature of u is given by $\kappa_g(t) = g(\nabla e_1(t)/dt, e_2(t))$ as in (4.2.6).

6.3.1 Proposition. *Under the above conditions, there exists a differentiable function $\theta(t)$, $t \in I$, such that*

(1) $$e_1(t) = \cos\theta(t) \cdot E_1(u(t)) + \sin\theta(t) E_2(u(t)).$$

The function $\theta(t)$ is uniquely determined up to integral multiples of 2π and satisfies

(2) $$\kappa_g(t) = \dot\theta(t) + \sqrt{g_{,1}}\dot{u}^2(t).$$

Remark. This proposition generalizes (1.4.1) and (2.1.3) (where the analogous result was proved for curves in Euclidean space) to curves on surfaces with a Riemannian metric. In the Euclidean case, we defined $\theta(t)$ with respect to a parallel translation invariant orthonormal 2-frame, namely e_1, e_2. Such a 2-frame *does not in general exist* on a surface with $K \not\equiv 0$ (see (4.4.2)).

PROOF. 1. The existence and uniqueness, modulo 2π, of $\theta(t)$ satisfying (1) follows by an argument identical to the one in (2.1.3).
2. We may write $e_2(t)$ as follows:

$$e_2(t) = -\sin\theta(t)E_1(u(t)) + \cos\theta(t)E_2(u(t)).$$

6.3 The Gauss–Bonnet Theorem

Since $g(E_i, E_k) = \delta_{ik}$, $g(\nabla E_i/dt, E_k) + g(E_i, \nabla E_k/dt) = 0$, so, in particular, $g(\nabla E_i/dt, E_i) = 0$. Therefore

$$\kappa_g(t) = \dot{\theta}(t) + g(\nabla E_1/dt, E_2)(t).$$

But $\nabla E_1/dt = \sum_{i,k} \dot{u}^i \Gamma^k_{1i} e_k$ and $\Gamma^2_{12} = \sqrt{g}_{,1}/\sqrt{g}$, $\Gamma^1_{11} = \Gamma^2_{11} = \Gamma^1_{21} = 0$ in geodesic coordinates (see (4.2.4)). From this, equation (2) follows. □

6.3.2 Theorem (Gauss–Bonnet, local version).[7] *Let M be an oriented surface with Riemannian metric. Suppose $P: F \to M$ is a diffeomorphism of a polygon F onto a subset of M. If α_j, $0 \le j \le k$, denote the exterior angles at the vertices of $P(F)$ and κ_g = the geodesic curvature of the boundary curve ∂P (traversed in the positive sense). Then*

(*) $$\iint_P K\, dM + \int_{\partial P} \kappa_g\, dt + \sum_j \alpha_j = 2\pi.$$

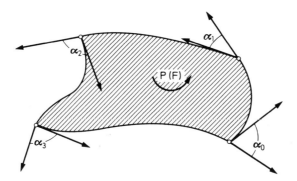

Figure 6.1 Gauss–Bonnet theorem

PROOF. 1. Suppose P lies entirely in one geodesic coordinate system (u^1, u^2). By (4.3.8), K may be written as div X:

$$K = -\frac{\sqrt{g}_{,11}}{\sqrt{g}} = \frac{1}{\sqrt{g}}\left\{\frac{\partial}{\partial u^1}\left(\sqrt{g}\left(-\frac{(\sqrt{g})_{,1}}{\sqrt{g}}\right)\right) + \frac{\partial}{\partial u^2} 0\right\},$$

where $X = (-\sqrt{g}_{,1}/\sqrt{g})e_1$. Using the divergence theorem (5.6.9),

$$\iint_P K\, dM = \int_{u \circ \partial P} \sqrt{g}\xi^1\, du^2 - \sqrt{g}\xi^2\, du^1 = -\int_{u \circ \partial P} \sqrt{g}_{,1}\, du^2.$$

We may parameterize ∂P to be a unit-speed, positively oriented, simply

---

[7] Bonnet, O. Memoire sur la théorie générale des surfaces. *J. de l'Ecole Polytechnique* **19**, H.32, 1–146 (1848). The important special case of a geodesic triangle (see (6.3.3 (ii))) was treated by Gauss in the "Disquisitiones."

6 The Global Geometry of Surfaces

closed curve $u(t) = (u^1(t), u^2(t))$, $t \in I$. Let $I_j = [a_j, b_j]$, $0 \leq j \leq k$, be subintervals on which $u_j = u \circ I_j$ is smooth. By (6.3.1),

$$-\int_{u \circ \partial P} \sqrt{g_{,1}}\, du^2 = \sum_j \left(\int_{I_j} \dot{\theta}(t)\, dt - \int_{I_j} \kappa_g(t)\, dt \right).$$

We claim that $\sum_j \int_{I_j} \dot{\theta}(t)\, dt + \sum_j \alpha_j = 2\pi$, which will prove the theorem in this special case.

2. *Proof of claim.* If the metric g were the Euclidean metric, i.e., if $g = g_{22} = 1$, then the claim would be precisely the Umlaufsatz (2.2.1). We now reduce the general case to the Euclidean case as follows. On U, let

$$ds_\tau^2 = (du^1)^2 + g_{\tau 22}(du^2)^2, \qquad 0 \leq \tau \leq 1,$$

be a family of line elements with

$$g_{\tau 22} = \tau + (1 - \tau) g_{22}.$$

For $\tau = 0$, ds_0^2 is the given line element on U and, for $\tau = 1$, ds_1^2 is the Euclidean line element. Notice that each ds_τ^2, $0 \leq \tau \leq 1$, is in fact a line element since $g_{\tau 22}$ is always strictly positive. For any $\tau \in [0, 1]$, we can define the exterior angles $\alpha_{\tau j}$ and the functions $\theta_{\tau j}$ as above. These functions will be continuous in τ, for

$$\cos \theta_\tau(t) = \frac{g_\tau(E_{\tau 1}, \dot{u})}{\sqrt{g_\tau(\dot{u}, \dot{u})}}, \quad \text{where } E_{\tau 1} := e_1; \quad E_{\tau 2} := e_2/\sqrt{g_{\tau 22}}$$

$$\sin \theta_\tau(t) = \frac{g_\tau(E_{\tau 2}, \dot{u})}{\sqrt{g_\tau(\dot{u}, \dot{u})}}$$

$$\cos \alpha_{\tau j} = \frac{g_\tau(\dot{u}(a_j), \dot{u}(b_{j-1}))}{\sqrt{g_\tau(\dot{u}(a_j), \dot{u}(a_j))} \cdot \sqrt{g_\tau(\dot{u}(b_{j-1}), \dot{u}(b_{j-1}))}}.$$

Furthermore, for every τ the number

$$2\pi n_\tau := \sum_j \int_{I_j} \dot{\theta}_\tau(t)\, dt + \sum_j \alpha_{\tau j} = \sum_j (\theta_\tau(b_{j-1}) - \theta_\tau(a_j)) + \alpha_{\tau j}$$

is a multiple of 2π. Thus n_τ must be a constant *integer*, since it depends continuously on τ: $n_0 = n_\tau = n_1 = 2\pi$, since $2\pi n_1 = 2\pi$. This proves the claim.

3. We now remove the restriction that $P: F \to M$ has values lying inside of a single geodesic coordinate system. Given $P: F \to M$, we may subdivide F into $\{F_\rho\}$, $1 \leq \rho \leq f$, so that each F_ρ is a polygon and $P_\rho = P|F_\rho$ has values lying in some geodesic coordinate system. For each ρ we have

(*) $$\iint_{P_\rho} K\, dM + \int_{\partial P_\rho} \kappa_g\, dt = 2\pi + \sum_{j_\rho} (\beta_{j_\rho} - \pi),$$

where the sum on the right is taken over all the vertices of P_ρ and $\alpha_{j_\rho} = \pi - \beta_{j_\rho}$.

Denote the number of vertices of the subdivision $\{F_\rho\}$ by v. Denote the number of edges by e and the number of faces or surfaces by f. Then

$v - e + f = 1$. This can be proved as follows: If each F_ρ is a triangle, it follows from induction on the number of triangles since adjoining a triangle to a *triangulation* (i.e., a subdivision by triangles) does not change the sum $f - e + v = 1$. Given a general subdivision into polygons, refining it to a triangulation does not change the sum $f - e + v$ (proof by induction on f).

Summing over ρ, the left-hand side becomes $\iint_P K\, dM + \int_{\partial P} \kappa_g\, dt$ since the inner edges are each traversed twice, once in each direction, and thus cancel out. The right-hand side may be computed as follows. First, $2\pi f \cdot \sum_\rho \sum_{j_\rho} \beta_{j_\rho} = \sum_j \beta_j + 2\pi \mathring{v}$, where $\mathring{v}$ is the sum of the inner vertices and $\sum_j$ is taken over the exterior vertices, i.e., vertices of F. Now, $\sum_\rho \sum_{j_\rho}(-\pi) = -2\pi\mathring{e} + \sum_j(-\pi)$, where $\mathring{e}$ is the number of internal edges. Since $-\mathring{e} + \mathring{v} = -e + v$, the right-hand side is thus equal to $2\pi(f - e + v) + \sum_j(\beta_j - \pi) = 2\pi - \sum_j \alpha_j$. □

6.3.3 Corollaries. i) If $\beta_j := \pi - \alpha_j$ are the interior angles at the k corners of the polygon P, then

$$\iint_P K\, dM + \int_{\partial P} \kappa_g\, dt = \sum_j \beta_j + (2 - k)\pi.$$

ii) (Gauss' theorema elegantissimum). *If the k edges of the polygon P are geodesics ($\kappa_g = 0$), then $\iint_P K\, dM = \sum_j \beta_j + (2 - k)\pi$. In particular, for $k = 3$ (a geodesic triangle):*

$$\sum_j \beta_j = \pi + \iint_P K\, dM.$$

iii) *Suppose $K = K_0 = $ constant and the edges of P are geodesics. Let $A(P) = \iint_P dM$ be the area of P. Then $\sum_j \beta_j = (k - 2)\pi + K_0 A(P)$. If, in addition, P is a triangle, then $\sum_j \beta_j = \pi + K_0 \cdot A(P) \geq 0$. In words, the sum of the interior angles of a geodesic triangle on a surface of constant curvature K_0 is equal to π plus K_0 times the area of the interior of the triangle. If $K_0 < 0$, then $A(P) \leq -\pi/K_0$.*

iv) *If $K \leq 0$, then there cannot exist a geodesic 2-gon, since that would mean $\sum \beta_j \leq 0$, a contradiction.*

Theorem (6.3.2) has some very important applications to the theory of compact orientable surfaces with a Riemannian metric, namely the relationship between $\iint_M K\, dM$ and the Euler characteristic of M, which we now define.

6.3.4 Definitions. Suppose M is a differentiable orientable compact surface. Let $\Pi := \{P_\rho: F_\rho \to M \mid 1 \leq \rho \leq f\}$ be a polygonal subdivision as defined in (5.6.10). Let v be the number of vertices of Π (that is the sum of the points of M which are the images of the vertices of some F_ρ). Let e be the sum of the edges of Π, defined similarly, and let f be the sum of the faces of Π. The *Euler characteristic of M (with respect to Π)* is the number

$$\chi_\Pi(M) = f - e + v.$$

A polygon $P_\rho: F \to M$ is orientation-preserving if, for any positively oriented chart (u_α, M_α),

$$u_\alpha \circ P_\rho: F_\rho \cap P_\rho^{-1}(M_\alpha) \subset \mathbb{R}^2 \to U_\alpha \subset \mathbb{R}^2$$

is orientation-preserving.

In part 3 of the proof of (6.3.2) the sum $f - e + v$ (with respect to a subdivision of a polygon F) was introduced and it was shown that $f - e + v$ is always equal to $+1$. The Euler characteristic $\chi_\Pi(M)$ is a generalization of this number to polygonal subdivisions of compact orientable surfaces, M. As the proof of the previous theorem shows, $\chi_\Pi(M)$ remains unchanged by a refinement of Π. Thus we may assume, without loss of generality, that in our definition of $\chi_\Pi(M)$ the polygons of Π are all triangles (or, if need be, quadrilaterals).

EXAMPLES. 1. $M = S^2$, the sphere. The polygonal subdivision Π of S^2 defined by projection onto an inscribed tetrahedron allows us to compute $\chi_\Pi(S^2) = 4 - 6 + 4 = 2$.

2. $M = T^2$, the torus, may be subdivided by using three meridians and three parallel curves. The resulting polygonal subdivision consists of quadrilaterals with a total of $f = 9$ faces, $e = 18$ edges, and $v = 9$ vertices. Thus $\chi_\Pi(T^2) = 0$.

3. Let M_0 be a compact oriented surface and let χ_{Π_0} be a polygonal subdivision of M_0. We may assume that Π_0 contains a quadrilateral, say P_0, introducing it if necessary by a subdivision of one of the polygons of Π_0. This will *not* alter $\chi_{\Pi_0}(M_0)$. It is possible to construct a new surface $M = M_0 + H$ by a process known as "attaching a handle H." Consider the torus with the quadrilateral subdivision Π_0' defined in (2) above. Let one of the quadrilaterals of Π_0' be labelled P_0'. Then $M_0 - P_0$ and $H = T^2 - P_0'$ both have boundaries which consist of four smooth curves which we may identify (see Figure 6.3). The resulting surface M inherits a polygonal subdivision, Π, equal to the union of $\Pi_0 - \{P_0\}$ and $\Pi_0' - \{P_0'\}$. Moreover, $\chi_\Pi(M) = \chi_{\Pi_0}(M_0) - 2$. This is because we have deleted two faces, four edges, and four vertices from $\Pi_0 \cup \Pi_0'$.

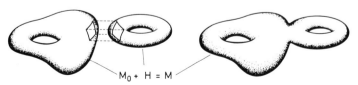

$M_0 + H = M$

Figure 6.2 Attaching a handle

6.3 The Gauss–Bonnet Theorem

Definition. A *surface of genus g* is a surface which is diffeomorphic to S^2 with $g \geq 0$ handles attached. By the above construction, such a surface has a polygonal subdivision Π with $\chi_\Pi(M) = 2 - 2g$.

6.3.5 Theorem (The Gauss–Bonnet theorem for closed surfaces). *Let M be a compact oriented surface with a Riemannian metric and let π be a polyhedral subdivision of M. Then $\iint_M K\, dM = 2\pi \cdot \chi_\Pi(M)$.*

6.3.6 Corollary. *The Euler characteristic of M is independent of the choice of polyhedral subdivision of M or the orientation of M.*

PROOF. 1. We proceed exactly as in part 3 of the proof of Theorem (6.3.2). First of all, formula (*) is valid for each ρ. Summing over ρ, the left-hand side becomes $\iint_M K\, dM$; all of the edges cancel pairwise since each one appears twice, with opposite orientation. The right-hand side adds up to $2\pi \cdot \chi_\Pi(M)$ because we have 2π for each face, $-\pi$ for each time an edge ends in a vertex (twice), and $\sum_{j_\rho} \beta_{j_\rho}$ equals 2π times the number of vertices. This proves the theorem.

2. To prove the corollary, simply observe that the left-hand side of the equation (*) depends only on M and its Riemannian metric, while the right-hand side is defined in terms of numbers which are independent of the orientation of M. $\square$

6.3.7 Theorem. *Suppose M is a compact orientable surface with a Riemannian metric.*
   i) *If $\chi(M) \geq 0$ (resp. > 0), then there exists a $p \in M$ with $K(p) \geq 0$ (resp. > 0).*
   ii) *If $\chi(M) \leq 0$ (resp. < 0), then there exists a $p \in M$ with $K(p) \leq 0$ (resp. < 0).*
   iii) *If $K > 0$, then $\chi(M) > 0$. [More precisely, $\chi(M) = 2$, for this is the only possible positive value of the Euler characteristic.] This implies that M is diffeomorphic to S^2.*
   iv) *If $K = 0$, then $\chi(M) = 0$. This implies that M is diffeomorphic to T^2.*
   v) *If $K < 0$, then $\chi(M) < 0$. [More precisely, M is a sphere with two or more handles.]*

The proof [with the exception of the bracketed statements] follows directly from (6.3.5). [The bracketed statements follow from the classification theorem for compact orientable surfaces. Namely, any such surface is diffeomorphic to a sphere with $g \geq 0$ handles (see Massey, *loc. cit.*).]

We end this section with an interesting application of the Gauss–Bonnet theorem to the theory of curves in $\mathbb{R}^3$.

6.3.8 Theorem (Jacobi).[8] *Suppose $c(t)$, $0 \leq t \leq \omega$, is a regular closed curve*

---
[8] Jacobi, C. G. J. Über einige merkwürdige Curventheoreme. *Schumacher's Astronomische Nachr.* **20**, Nr. 463, 115–120 (1842).

6 The Global Geometry of Surfaces

in $\mathbb{R}^3$ on which $\dot{c}(t)$ and $\ddot{c}(t)$ are linearly independent. Let $(e_1(t), e_2(t), e_3(t))$ be the unique Frenet frame of c. Suppose the closed curve $e_2(t)$, $0 \le t \le \omega$, which lies on S^2 is simple (i.e., without self-intersections). Then this curve divides S^2 into two sets of equal area $= 2\pi$.

PROOF. We may assume, without loss of generality, that t is arc length on $c(t)$. Define $r(t)$ by

$$\cos r(t) = \frac{\kappa}{\sqrt{\kappa^2 + \tau^2}}(t), \qquad \sin r(t) = \frac{\tau}{\sqrt{\kappa^2 + \tau^2}}(t).$$

Since $\dot{e}_2(t) = -\kappa(t)e_1(t) + \tau(t)e_3(t)$ and $\dot{e}_2(t) \cdot e_2(t) = 0$, the vector fields

$$E_1(t) := (-\cos r(t))e_1(t) + (\sin r(t))e_3(t)$$

and

$$E_2(t) := (\sin r(t))e_1(t) + (\cos r(t))e_3(t)$$

are the unit tangent and the unit normal vector fields on $e_2(t)$, respectively. This means that $(E_1(t), E_2(t))$ is the Frenet frame of $e_2(t)$ on S^2 since $(e_1(t), e_3(t))$ span $T_{e_2(t)}S^2$.

Thus

$$\frac{dE_1}{dt} = \dot{r}E_2(t) - \sqrt{\kappa^2 + \tau^2}\, e_2(t),$$

which implies

$$\frac{\nabla E_1}{dt} = \dot{r}E_2(t), \qquad \kappa_g(t) = \frac{\nabla E_1}{dt} \cdot E_2(t) = \dot{r}(t).$$

Suppose P is one of the connected subsets of S^2 bounded by $e_2(t)$, $0 \le t \le \omega$. By the Jordan curve theorem, P is a "polygon." Since $K = 1$ on S^2, Theorem (6.3.2) implies

$$\iint_P 1\, dM + \int_{\partial P} \dot{r}\, dt = \iint_P dM = 2\pi. \qquad \square$$

6.4 Completeness

In this section M will always be assumed to be a connected Riemannian manifold. When M is required to have dimension $= 2$, i.e., when M is a surface, this will be indicated.

6.4.1 Definition. The *distance* $d(p, q)$ between two points p and q in M is the infimum of the length $L(c)$ of all piecewise smooth curves c which join p to q.

We wish to show that $d(\ ,\ )$ actually defines a *metric* on M in the usual sense. In other words,

i) $d(p, q) \ge 0$ (equality $\Leftrightarrow p = q$);
ii) $d(p, q) = d(q, p)$; and
iii) $d(p, q) + d(q, r) \ge d(p, r)$ (triangle inequality).

6.4.2 Theorem. *The distance function $d(p, q)$ defines a metric on M. Moreover, the metric topology is equivalent to the topology of M.*

PROOF. 1. Certainly $d(p, q) \geq 0$, $d(p, p) = 0$, and $d(p, q) = d(q, p)$. Also the triangle inequality follows easily from the definition of $d(\ ,\ )$.

2. Suppose $d(p, q) = 0$. Consider a geodesic disk $B_\rho(p)$ centered at p. By (5.3.4), $d(p, q) > 0$ for all $q \notin B_\rho(p)$ and for any $q \in B_\rho(p)$, $d(p, q) \geq 0$ with equality if and only if $p = q$. Actually, only smooth curves are considered in the proof of (5.3.4). But piecewise smooth curves may also be admitted. One uses the fact that geodesic (polar) coordinates (u^1, u^2) have the characteristic property that *any* curve connecting (u_0^1, u_0^2) to (u_1^1, u_1^2) must have length at least $|u_1^1 - u_0^1|$. This is because the distance between orthogonal trajectories to the "$u^1 =$ constant" curves are given by the difference in the parameter values of these trajectories. This completes the proof that d is a metric.

3. A basis for the open sets in the metric topology consists of embedded geodesic disks $B_\rho(p)$, $\rho > 0$, $p \in M$. These we know are open sets in the usual topology. Conversely, given a neighborhood $U(p)$ of p, there exists a $\rho > 0$ with $B_\rho(p) \subset U(p)$. □

We know from Chapter 5, section 3, that for $\rho > 0$ sufficiently small $B_\rho(p)$ is a geodesic ρ-disk. (Recall that a geodesic disk is the image of $B_\rho(0) \subset T_pM$ under $\exp_p$ on which $\exp_p | B_\rho(0)$ is a diffeomorphism.) The maximal radius $\rho_m(p)$ such that $B_\rho(p)$ is a geodesic disk for all $\rho < \rho_m(p)$ is in general a function of p and *cannot* be explicitly computed from knowledge of the curvature of M alone. The number $\rho_m(p)$ is called the *radius of injectivity* at p. We know that $\rho_m(p)$ is positive, but it may be arbitrarily small as shown by the example of the pseudosphere (3.9.1(iii)). Likewise, $\rho_m(p)$ may be equal to $+\infty$. This happens for any point in Euclidean n-space.

By Theorem (5.2.5), for every $p \in M$, there exists a neighborhood $M(p)$ of p and a $\rho = \rho(p) > 0$ such that, for every $q \in M(p)$, $B_\rho(q)$ is a geodesic disk. If $K \subset M$ is a compact set, then there exists a finite set $\{p_i\} \subset K$ such that $\{M(p_i)\}$ covers K. Therefore, if $\rho < \rho(K) = \min_i \{\rho(p_i)\}$, $B_\rho(q)$ is a geodesic disk for all $q \in K$. We rewrite this result as follows.

6.4.3 Proposition. *Let K be a compact set in M, a surface with Riemannian metric. Then there exists a number $\rho = \rho(K) > 0$, depending only on K, such that, for all $p \in K$, $\exp_p | B_\rho(0) \colon B_\rho(0) \to M$ is an injective diffeomorphism: $B_\rho(p) = \exp_p B_\rho(0)$ is an embedded geodesic disk of radius ρ.*

When we defined the exponential map in (5.2), its domain of definition was a suitably small neighborhood of the zero vectors in TM. The objects of interest in Riemannian geometry in the large are those surfaces or manifolds M for which $\exp_p$ is defined on all of T_pM.

6.4.4 Definition. *A surface (or manifold) is said to be geodesically complete if the exponential map is defined on all of TM.*

An important theorem of Hopf and Rinow[9] characterizes geodesic completeness in several ways. Among other things, it states that M is geodesically complete if and only if M is complete in the metric $d(\ ,\ )$ defined in (6.4.1). (A metric space is complete if and only if every Cauchy sequence converges.) We will not prove this completely, but content ourselves with proving half of the equivalence.

6.4.5 Lemma. *Suppose M is complete as a metric space. Then M is geodesically complete.*

Note: The hypothesis is certainly satisfied if M is compact.

PROOF. 1. Let $X \in T_p M$ be a unit vector. We wish to show that the geodesic $c_X(t) = \exp_p tX$ is defined for all $t \in \mathbb{R}^+ = \{t \in \mathbb{R} \mid t \geq 0\}$. We know that $c_X(t)$ is defined for an interval of the form $[0, t^*[$. Let $\{t_n\}$ be a sequence in $[0, t^*[$ with $\lim_n t_n = t^*$. Since $d(\exp_p t_k X, \exp_p t_l X) \leq |t_k - t_l|$, $\{p_n = \exp_p t_n X\}$ is a Cauchy sequence. The assumption that M is metrically complete implies that there exists a $q \in M$ with $\lim_n p_n = q$.

2. According to (5.2.5), there exists a neighborhood M_0 of q and a $\rho > 0$ such that for every $p^* \in M_0$ the exponential map $\exp_p$ is defined on $B_\rho(0) \subset T_p M$.

By choosing n large enough to make $t^* - t_n < \rho/2$, we may insure that $p_n \in M_0$. This means that the geodesic ray emanating from $c_X(t_n)$ with initial direction $\dot{c}_X(t_n) \in T_{p_n} M$ is defined for all $|t| < \rho$, so that $c_X(t)$ is defined for $t \in [0, t_n + \rho)$. But $t_n + \rho > t^*$, and thus $c_X(t)$ is defined for all $t \geq 0$. □

The most important property of geodesically complete surfaces and manifolds is contained in the following theorem.

6.4.6 Theorem (Hopf–Rinow).[9] *Suppose M is geodesically complete and connected. Then any two points of M may be joined by a minimal geodesic whose length is equal to $d(p, q)$.*

Note: For the definition of "minimal geodesic," see (5.3.3).

Before proving the theorem, the reader is urged to notice that the converse of the theorem is *not true*. For example, the interior of the unit-disk of $\mathbb{R}^2$ with the Euclidean metric satisfies the conclusion of the theorem (any two points may be joined by a straight line (minimal geodesic) lying inside the disk), but is not complete.

PROOF. 1. Without loss of generality, we may assume that $d(p, q) = r > 0$. Let ρ be such that $0 < \rho < r$ and $\exp_p|B_\rho(0)$ is a diffeomorphism from $B_\rho(0) \subset T_p M$ to $B_\rho(p)$. Choose ϵ satisfying $0 < \epsilon < \rho$, and define $S = S_\epsilon(p)$ to be equal to $\exp_p S_\epsilon(0)$, where $S_\epsilon(0)$ is the hypersphere of radius ϵ, centered at $0 \in T_p M$.

[9] Hopf, H., and Rinow, W. Über den Begriff der vollständigen differentialgeometrischen Flächen. *Math. Ann.* **116**, 749–766 (1938).

6.4 Completeness

Since S is compact, there exists a $p_0 \in S$ such that $d(p_0, q) \leq d(p', q)$ for all $p' \in S$. Let $X \in T_pM$ be the unique unit tangent vector such that $p_0 = \exp_p \epsilon X$. We will show that $\exp_p rX = q$ and thus $c(t) = \exp_p tX$, $0 \leq t \leq r$, is a minimal geodesic from p to q.

2. Toward that end, we shall prove that for $t \in [0, r]$,

$((t))$ $\qquad\qquad d(c(t), q) = r - t.$

We know that $((t))$ holds for $t = \epsilon$. Since every curve from p to q must pass through S,

$$r = d(p, q) = \min_{p' \in S}(d(p, p') + d(p', q)) = \epsilon + d(p_0, q) = \epsilon + d(c(\epsilon), q),$$

which implies $((\epsilon))$. Similarly, $((t))$ holds for all $t \leq \epsilon$.

Suppose now that $t_0 \in [0, r]$ is the supremum of all t' such that $((t))$ holds for $t \in [0, t'[$. By the paragraph above, $t_0 \geq \epsilon$. By continuity of both sides of the equation $d(c(t), q) = r - t$, it follows that $((t_0))$ holds.

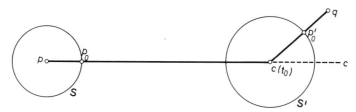

Figure 6.3   Construction of a minimal geodesic

Suppose that $t_0 < r$. We will arrive at a contradiction. Let S' be a small hypersphere centered at $c(t_0)$ with radius ϵ', $0 < \epsilon' < r - t_0$. If $p'_0 \in S'$ is a point on S' whose distance from q is the minimum for all points on S', and $c'(t)$, $t_0 \leq t \leq t_0 + \epsilon'$, is the minimal geodesic from $c(t_0)$ to p'_0, then

$$d(c(t_0), q) = \min_{q' \in S'}(d(c(t_0), q') + d(q', q)) = \epsilon' + d(p'_0, q),$$

i.e.,

(*) $\qquad\qquad d(p'_0, q) = (r - t_0) - \epsilon'.$

But $p'_0 = c(t_0 + \epsilon)$. To prove this, first observe that

$$d(p, p'_0) \geq d(p, q) - d(p'_0, q) = r - (r - t_0) + \epsilon' = t_0 + \epsilon'.$$

But since the composite curve $c \mid [0, t_0]$ followed by c' has length $t_0 + \epsilon' \leq d(p, p'_0)$, it follows that it is an *unbroken* geodesic, i.e., $p'_0 = c'(\epsilon) = c(t_0 + \epsilon)$. The relation (*) now implies $((t_0 + \epsilon'))$, contradicting the definition of t_0. Therefore $t_0 = r$ and $((r))$ is our claim. □

Remarks. 1. The careful reader is encouraged to pinpoint exactly where in the proof the hypothesis of geodesic completeness was used.

2. Minimal geodesic joins between two points need not be unique. The simplest example is the sphere on which any two antipodal points may be joined by uncountably many minimal geodesics.
3. In the special case that M is simply connected and the curvature is non-positive, a strengthened version of Theorem (6.4.6) will be proved, albeit in a quite different way (see (6.6.4)).

6.5 Conjugate Points and Curvature

In this section M will always denote a complete surface with a Riemannian metric. The first few results obtained may be generalized to complete Riemannian manifolds, with little or no changes in the proofs. The comparison theorems are somewhat harder in the general case.

We shall have need to refer to section 5.4, which provides some basic results concerning Jacobi fields.

6.5.1 Definition. Let $c = c(t)$, $t \geq 0$, be a geodesic ray on M with $c(0) = p$ and $\dot{c}(0) \neq 0$. Let $\tilde{c}(t) = t\dot{c}(0)$, $t \geq 0$, be the ray in T_pM for which $\exp_p \tilde{c}(t) = c(t)$. A point $c(t_1)$, $t_1 \geq 0$, is said to be *conjugate* to $p = c(0)$ (along $c \mid [0, t_1]$) provided

$$d(\exp_p)_{\tilde{c}(t_1)} : T_{\tilde{c}(t_1)}(T_pM) \to T_{c(t_1)}M$$

is not bijective, i.e., $\exp_p : T_pM \to M$ is not regular at $\tilde{c}(t_1)$.

Remarks. 1. A conjugate point of $c(0)$ along c can only occur for some $t_1 > 0$, since $(d \exp_p)_0$ is bijective (see (5.2.4)).
2. Since $d(\exp_p)_{\tilde{c}(t)} \dot{\tilde{c}}(t) = \dot{c}(t) \neq 0$, the kernel of the linear map $(d \exp_p)_{\tilde{c}(t)}$ is always in the complement of the one-dimensional linear subspace of $T_pM_{\tilde{c}(t)}$ determined by $\dot{\tilde{c}}(t)$. In fact, the proof of the next proposition will imply that the kernel is orthogonal to the line spanned by $\dot{\tilde{c}}(t)$.

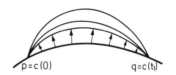

Figure 6.4 A conjugate point

6.5.2 Proposition. *The following statements are equivalent:*
   i) *$q = c(t_1)$ is conjugate to $p = c(0)$ along $c \mid [0, t_1]$.*
   ii) *There exists a nontrivial Jacobi field $Y(t)$ along $c(t)$, $0 \leq t \leq t_1$, $t_1 > 0$, with $Y(0) = Y(t_1) = 0$.*

PROOF. Using (5.4.3), we may assert the existence, for t sufficiently small, of a nontrivial Jacobi field $Y(t)$ with $Y(0) = 0$ and $A = (\nabla Y/dt)(0) \neq 0$, where

6.5 Conjugate Points and Curvature

A is orthogonal to $\dot{c}(0)$. In fact, $Y(t)$ may be written in the form

$$Y(t) = (d \exp_p)_{\tilde{c}(t)} tA.$$

This expression for a Jacobi field is valid for arbitrarily large t. The proof of (5.4.3) carries over verbatim to the case of a geodesic c, defined on an arbitrary nonempty open interval $I \subset \mathbb{R}$, where $c(I)$ lies within some coordinate chart (u, M_0) of M.

Thus $\hat{A} \in T_{\tilde{c}(t_1)}(T_p M)$ is a nonzero element of the kernel of $(d \exp_p)_{\tilde{c}(t_1)}$ if and only if the Jacobi field $Y(t)$ that is determined by $A = \hat{A}/t_1$ satisfies $Y(t_1) = 0$. This proves the proposition. □

We are now in a position to generalize the results of Theorems (4.3.9) and (5.3.4) about the length-measuring properties of geodesics.

6.5.3 Theorem. *Suppose $c = c(t)$, $0 \leq t \leq a$, $a > 0$, is a unit-speed geodesic on M which contains no conjugate points, i.e., no point of c is conjugate to $c(0)$ along c. Then for any curve b which is sufficiently close to c and joins $c(0)$ to $c(a)$, $L(b) \geq L(c)$.*

PROOF. Consider the differentiable function $\phi: \,]-\delta, a+\delta[\,\times\,]-\epsilon, \epsilon[\,\to\, M$ defined by

$$(r, \theta) \mapsto \exp_p((r \cos \theta)e_1(p) + (r \sin \theta)e_2(p)),$$

where $\{e_1(p), e_2(p)\} = \{\dot{c}(0), A\}$ is an orthonormal basis of $T_p M$. This function was introduced in the proof of (5.4.3), where the existence and regularity (for $r > 0$) of ϕ is proved for sufficiently small $\epsilon > 0$, $\delta > 0$. Locally, the map $\phi|\{r > 0\}$ is a coordinate map. In fact for $r > 0$, ϕ defines (locally) polar coordinates which, by (5.3.2), are geodesic coordinates based upon an arc of the geodesic circle $\exp_p\{S_r(0)\}$. Suppose $b = b(s)$, $s_0 \leq s \leq s_1$, is a curve from $c(0) = b(s_0)$ to $c(a) = b(s_1)$ which is sufficiently close to c to lie within the range of ϕ. As in (4.3.9), the length of b between parallel curves $u^1 = r = r_1 = $ constant and $u' = r = r_2 = $ constant is equal to or greater than $|r_2 - r_1|$ = distance between these parallel curves = length of $c(t)$ between these curves. Therefore $L(b) \geq L(c)$. □

Remark. This theorem has the following partial converse: Suppose c contains a point in its interior which is conjugate to $c(0)$. Then in every neighborhood of c there exists a curve b joining $c(0)$ to $c(a)$ which is *strictly shorter* than c.

The proof of this result uses the second variation formula for the length integral, and, while not difficult, is long, and we prefer to omit it.[10]

In the limit case, where the end-point $c(a)$ is conjugate to $c(0)$ along c, it is not possible to say in general whether c is locally the shortest curve from $c(0)$ to $c(a)$ or not. The situation is rather like the case of a real valued function

---

[10] For a proof of this result, see Gromoll–Klingenberg–Meyer [A6] or Kobayashi, S. On conjugate and cut loci. In: *Studies in Global Geometry and Analysis* [B9], or Bishop and Crittenden [B2].

6 The Global Geometry of Surfaces

$f(x)$ with $f'(x_0) = f''(x_0) = 0$. The function f may or may not have a local minimum at x_0.

6.5.4 Examples of Jacobi fields. On a surface of constant curvature $K = K_0$ the differential equation for a Jacobi field is $\ddot{y}(t) + K_0 y(t) = 0$ (see (5.4.1)). Actually the Jacobi field is $Y(t) = y(t)e_2(t)$, where $e_2(t)$ is a unit normal vector field along the geodesic in question. We are interested in solutions with $y(0) = 0$, $\dot{y}(0) = a \neq 0$.

$$y(t) = a \sin(t \cdot \sqrt{K_0}), \quad \text{if } K_0 > 0,$$
$$y(t) = at, \quad \text{if } K_0 = 0,$$
$$y(t) = a \sinh(t \cdot \sqrt{|K_0|}), \quad \text{if } K_0 < 0.$$

Thus conjugate points occur only in the case $K_0 > 0$, since the functions at and $a \sinh(t \cdot \sqrt{|K_0|})$ have no zeros when $t > 0$.

If the curvature of M is not constant it is still possible, under certain conditions, to obtain qualitative information about the occurrence of conjugate points. The main result we will prove along these lines is Theorem (6.5.6). To prove this theorem, we will need the following result from the theory of ordinary differential equations.

6.5.5 Lemma (Sturm comparison theorem).[11] *Let $u(t)$ be a solution to $\ddot{u}(t) + A(t)u(t) = 0$ with $u(0) = 0$, $\dot{u}(0) = 1$, and $v(t)$ a solution to $\ddot{v}(t) + B(t)v(t) = 0$ with $v(0) = 0$, $\dot{v}(0) = 1$. Suppose $A(t) \geq B(t)$. If a and b are the first zeros, after $t = 0$, of $u(t)$ and $v(t)$, respectively, then $a \leq b$. Furthermore, for t_0, t_1 satisfying $0 < t_0 < t_1 < a$, $v(t_1)u(t_0) \geq u(t_1)v(t_0)$ and $v(t_1) \geq u(t_1)$.*

(If $A(t) > B(t)$, then $a < b$, $v_1(t_1)u(t_0) > u(t_1)v(t_0)$, and $v(t_1) > u(t_1)$.)

PROOF. 1. Since $\dot{u}(0) = \dot{v}(0) = 1$, $u(t) > 0$ for all t, $0 < t < a$, and $v(t) > 0$ for all t, $0 < t < b$. Assume that $a > b$. We have

$$0 = \int_0^b u(\ddot{v} + Bv) - v(\ddot{u} + Au) \, dt = (u\dot{v} - v\dot{u})\big|_0^b + \int_0^b (B - A)uv \, dt.$$

Since $A(t) \geq B(t)$, the integrand $(B - A)uv$ on the right is nonpositive in the interval $[0, b]$, which means that $u\dot{v} - v\dot{u}\big|_0^b = u(b)\dot{v}(b)$ is nonnegative. But $u(b) > 0$ and $\dot{v}(b) < 0$, a contradiction.

2. Suppose $0 < t < a$. Since

$$0 = \int_0^t u(\ddot{v} + Bv) - v(\ddot{u} + Au) \, dt$$
$$= (u\dot{v} - v\dot{u})\big|_0^t + \int_0^t (B - A) uv \, dt \leq (u\dot{v} - v\dot{u})\big|_0^t,$$

---

[11] Sturm, J. C. F. Mémoire sur les équations différentielles du second ordre. *J. Math. Pures Appl.* **1**, 106–186 (1836).

$(d/dt)(\log v(t)) \geq (d/dt)(\log u(t))$. Thus if $0 < t_0 \leq t_1 < a$, $v(t_1)u(t_0) \geq u(t_1)v(t_0)$. Now
$$\lim_{t_0 \to 0} v(t_0)/u(t_0) = 1 \quad \text{and} \quad u(0) = v(0) = 0$$
imply that $v(t_1) \geq u(t_1)$.

3. If $A(t) > B(t)$, an analogous proof gives the sharper results. □

6.5.6 Theorem. *Suppose $c(t)$, $t \geq 0$, is a unit-speed geodesic. Define $K(t) = K \circ c(t)$.*
   i) If $K(t) \leq K_1$ for all t, then $c(0)$ has no conjugate points along c for $t \in [0, \pi/\sqrt{K_1}[$. (If $K(t) < K_1$, $c(0)$ has no conjugate points along c for $t \in [0, \pi/\sqrt{K_1}]$.)
   ii) If $0 < K_0 \leq K(t)$ for all t, then $c(t)$ must have at least one conjugate point in $]0, \pi/\sqrt{K_0}]$. (If $K_0 < K(t)$, then $c(0)$ must have a conjugate point in $]0, \pi/\sqrt{K_0}[$.)
   In case $K' \leq 0$, we interpret $\pi/\sqrt{K'}$ to be $+\infty$.

PROOF. 1. Suppose $K(t) \leq K_1$ and assume $K_1 > 0$. Let $B(t) = K(t)$, $A(t) = K_1$, and apply (6.5.5) above. The solution $u(t)$ is equal to $\sin(t\sqrt{K_1})/\sqrt{K_1}$. Thus $v(t)$ cannot vanish for $t < \pi/\sqrt{K_1}$, which means that any nontrivial Jacobi field along $c(t)$ with initial value $= 0$ cannot have another zero in $]0, \pi/\sqrt{K_1}[$. By Proposition (6.5.2), $c(t)$ has no conjugate points in $]0, \pi/\sqrt{K_1}[$.

2. The other cases of (i), as well as (ii) are proved analogously. (The sharp inequalities follow from the sharp inequalities of (6.5.5).) □

Remark. This result can be interpreted as a comparison theorem, comparing, qualitatively, the placement of conjugate points along geodesics on a surface of bounded curvature with the well-known distribution of conjugate points on an appropriate surface of constant curvature K_1 or K_0. See the examples in (6.5.4).

These examples also show that the inequalities in (6.5.6) are the best possible ones.

6.5.7 Corollary. *Suppose $c(t)$, $t \geq 0$, is a geodesic on which $K_0 \leq K \circ c(t) \leq K_1$. Then $c(0)$ has no conjugate points along c for $t \in [0, \pi/\sqrt{K_1}[$ and at least one conjugate point in $[\pi/\sqrt{K_1}, \pi/\sqrt{K_0}]$.*

Since we are assuming M to be complete in this section, it is worth noticing what the condition $K \geq K_0 > 0$ implies for complete surfaces. First of all, Theorem (6.5.6) implies that every geodesic segment of length greater than $\pi/\sqrt{K_0}$ has a conjugate point in its interior (with respect to the initial point). By the converse to (6.5.3), which we stated but did not prove, such a geodesic segment *cannot measure length*, so it is not a minimal geodesic connecting its end-points. Therefore we have the following theorem.

6.5.8 Theorem (Bonnet).[12] *On a complete surface M with $K \geq K_0 > 0$, the distance between any two points is at most $\pi/\sqrt{K_0}$. Therefore, M is a complete bounded metric space and hence is compact.*

As we pointed out at the beginning of this section, the results about conjugate points hold true for n-dimensional Riemannian manifolds. Bonnet's theorem also generalizes. The necessary curvature inequalities involve *sectional curvature*.

6.6 Curvature and the Global Geometry of a Surface

In this section, M will always be a complete surface with a Riemannian metric. The assumption that the Gauss curvature of a surface M lies in some predetermined interval has some important consequences for the geometry of M. The results of the previous section will play a central part in the discussion.

6.6.1 Theorem. *Suppose $K \leq K_1$ on M. Then a geodesic segment of length $< \pi/\sqrt{K_1}$ is the shortest curve joining its end-points when compared with all curves remaining in a sufficiently small neighborhood of the segment.*

This follows directly from (6.5.6) together with (6.5.3).

It is easy to see that if $K \leq K_1$, a geodesic of length $\pi/\sqrt{K_1}$ need not be a minimal geodesic joining its end-points: First of all, this could happen because M was not simply connected. For example, on the flat torus ($K \equiv 0$), there exist closed geodesics which can be considered as joining a point p to itself, and $d(p, p) = 0$.

A simply-connected counterexample may be constructed as follows: Consider a surface of revolution that looks like two globes of radius $= 1$, connected by a very narrow neck—an hourglass with a tapered waist. The curvature on the globular parts can be bounded above by a constant equal to 1, while the curvature of near the waist will be negative. Consequently, $K \leq K_1 = 1$ on this surface. However, the parallel circle at the waist will be a closed geodesic (by (4.5.1)). Since we may make the waist as small as we like, the closed geodesic can be made to have length strictly less than $\pi/\sqrt{K_1} = \pi$.

---

[12] Bonnet, O. Sur quelques propriétés des lignes géodésiques. *C.R. Acad. Sci. Paris* **40**, 1311–1313 (1855). Actually, Bonnet proved the following result: The "outer diameter" of an ovaloid (i.e., the maximum distance, in Euclidean space, between a pair of points on the ovaloid) is bounded above by $\pi/\sqrt{\min K}$. A proof of the theorem stated above may be found in Gromoll–Klingenberg–Meyer [B9] or Kobayashi–Nomizu [B13].

6.6 Curvature and the Global Geometry of a Surface

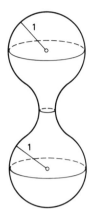

Figure 6.5 Hourglass with curvature $\leq K_1 = 1$

In this example, K takes on some strictly negative values (on the waist). On compact surfaces which are simply connected and satisfy $0 < K \leq K_1$, the conclusion of (6.6.1) holds even globally, i.e., a geodesic segment of length $\leq \pi/\sqrt{K}$ (not just $< \pi/\sqrt{K}$) is the shortest curve joining its end-points when compared with *all* curves on M. A compact simply-connected surface with $K > 0$ is isometric to a convex surface in Euclidean space, that surface being unique up to rigid motions of Euclidean space (see (6.8.1)).

Unfortunately, we cannot prove this result here.

See (6.8.3) for more discussion on this subject as well as references [A6] (the Kobayashi article), [B9], and [B13].

6.6.2 Lemma.[13] *Let $K \leq K_1$ on M. Define $\rho = \pi/\sqrt{K_1}$. Suppose $c = c(t)$, $0 \leq t \leq a$, is a unit-speed geodesic from $p = c(0)$ to $q = c(a)$, $a < \rho$. Let $b = b(s)$, $s_0 \leq s \leq s_1$, be another curve from p to q which may be written as*

$$b(s) = \exp_p \tilde{b}(s),$$

where $\tilde{b}(s)$ is a curve lying in $B_\rho(0) \subset T_pM$ with $\tilde{b}(s_0) = 0$, $\tilde{b}(s_1) = a\dot{c}(0)$. Then $L(b) \geq L(c)$.

Note: Compare this result with (5.3.4). There the conclusion is stronger, but the hypothesis is also stronger; ρ must be less than the injectivity radius at p.

PROOF. Since $K \leq K_1$, it follows from (6.5.6 (i)) and (6.5.1) that $\exp_p: B_\rho(0) \to M$ is a local diffeomorphism. By means of this diffeomorphism, the Riemannian metric g on M induces a Riemannian metric $\tilde{g}$ on $B_\rho(0)$:

$$\tilde{g}_{\tilde{p}}(\tilde{X}, \tilde{Y}) := g_{\exp_p \tilde{p}}(d \exp_p \tilde{X}, d \exp_p \tilde{Y}).$$

[13] The proof depends on the "Gauss lemma" which says that radial geodesics emanating from p cut geodesic circles (centered at p) orthogonally. This follows from (5.3.2). For a more general proof, which works for manifolds, see Gromoll–Klingenberg–Meyer [B9], p. 137, or Bishop–Crittenden [B2], p. 147.

6 The Global Geometry of Surfaces

With respect to this metric, $\exp_p | B_\rho(0)$ is a local isometry. Observe that polar coordinates in $B_\rho(0)$ are geodesic polar coordinates for the surface $(B_\rho(0), \tilde{g})$. The theorem now follows from (5.3.4) applied to $(B_\rho(0), \tilde{g})$. □

As the example of the hourglass surface with a narrow waist and curvature $K \le K_1 = 1$ shows, closed geodesics need not be "long" in the sense that no a priori lower bound on their length can be predicted from an upper bound on the Gauss curvature. However, the situation is not hopeless. Let us take a closer look at the hourglass example. Consider this waist geodesic, c, to be a geodesic segment whose initial point is equal to its end-point. Any family of curves which describes a deformation of the geodesic c into the trivial geodesic formed by the initial-point (=end-point) of c (keeping the initial-point-end-point fixed) contains curves which are "long" in the sense that

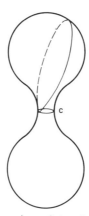

Figure 6.6 Deformation of the closed waist-geodesic

they will have length $\ge 2/\sqrt{K_1} = 2\pi$. This may be seen intuitively by looping a curve over one of the hemispheres. We will now make this precise. Suppose $c_0 = c_0(t), 0 \le t \le 1$, and $c_1 = c_1(t), 0 \le t \le 1$, are two curves from p to q. The curves c_0 and c_1 are said to be *homotopic* if there exists a continuous function $h: [0, 1] \times [0, 1] \to M$ such that each $c_s(t) = h(t, s), 0 \le t \le 1$, is a curve from p to q and $c_0(t) = h(t, 0)$ and $c_1(t) = h(t, 1)$ are the given initial curves. The family $c_s, 0 \le s \le 1$, is called a *homotopy* from c_0 to c_1.

6.6.3 Lemma (Klingenberg).[14] *Let c_0 and c_1 be two distinct geodesics from p to q with $L(c_0) \le L(c_1)$. Suppose $c_s, 0 \le s \le 1$, is a homotopy from c_0 to c_1. Then, if $K \le K_1$, there exists $s_0 \in [0, 1]$ such that*

$$L(c_{s_0}) + L(c_0) \ge 2\pi/\sqrt{K_1}.$$

[14] See Klingenberg, W. Über riemannsche Mannigfaltigkeiten mit positiver Krümmung. *Comment. Math. Helv.* **35**, 47–54 (1961).

6.6 Curvature and the Global Geometry of a Surface

Remarks. 1. Before proving this lemma, note its relevance to the preceding discussion. Let c_1 be the waist-geodesic and let c_0 be the constant "geodesic"; $c_0(t) \equiv p$. Then, since $L(c_0) = 0$, the lemma implies the above claim.
2. The inequality in the lemma is best possible as is shown by the example of antipodal points on a sphere connected by great semi-circles.
3. The lemma has interesting consequences in the case that $K_1 \leq 0$. Since we interpret $2\pi/\sqrt{K_1} = +\infty$ in this case, it means that two distinct geodesics from p to q cannot be homotopic. In particular, a closed geodesic cannot be homotopic to a constant curve. This fact will be exploited in Theorem (6.6.4) below.

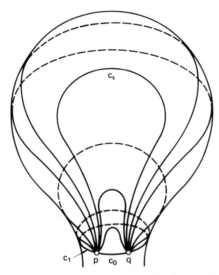

Figure 6.7 Homotopy (Adapted from Manfredo P. do Carmo, *Differential Geometry of Curves and Surfaces*, Prentice-Hall, Inc., 1976, p. 389.)

PROOF. Let $\pi/\sqrt{K_1} = \rho$. Since $K \leq K_1$, (6.5.6) implies that all geodesics emanating from p are free of conjugate points in $B_\rho(p)$. By (6.5.2), this means that $\exp_p B_\rho(0)$ is regular (i.e., of maximal rank). If $L(c_0) \geq \rho$ there is nothing to prove, so we might as well assume $L(c_0) < \rho$. Let $\tilde{c}_0 = \tilde{c}_0(t) = t\dot{c}(0)$, $0 \leq t \leq t_0$, be the line segment in T_pM which begins at $0 \in T_pM$, satisfies $c_0(t) = \exp_p \tilde{c}_0(t)$, and ends at $\tilde{q} = \tilde{c}_0(t_0) \in B_\rho(0)$.

For sufficiently small s, the curves c_s may be lifted to curves $\tilde{c}_s$ from 0 to $\tilde{q}$ which lie in $B_\rho(0)$, i.e., there exist curves $\tilde{c}_s$ such that $c_s(t) = \exp_p \tilde{c}_s(t)$. The curves $\tilde{c}_s(t)$ depend continuously on s. (Since $\exp_p | B_\rho(0)$ is a local diffeomorphism each $\tilde{c}_s$ must end at $\tilde{q}$.)

But such a lifting $\tilde{c}_s$ cannot exist for all $s \in [0, 1]$. For, since c_1 is a geodesic, this would force c_1 to be equal to c_0, contradicting the assumption that $c_0 \neq c_1$. Therefore to each $\epsilon > 0$ there must exist an $s = s(\epsilon) \in [0, 1[$ such that the curve $\tilde{c}_s \subset B_\rho(0)$, defined above, comes within distance ϵ of the boundary of $B_\rho(0)$. The length of the curve $\tilde{c}_s$ must be at least $2\rho - 2\epsilon - L(c_0)$. Consequently, by (6.6.2), $L(c_s) + L(c_0) \geq 2\rho - 2\epsilon$. But this inequality holds for all $\epsilon > 0$, and therefore the lemma follows. □

Remark. This lemma and its proof carry over word-for-word to Riemannian manifolds.

We will now use this lemma to prove a famous theorem of Hadamard.

6.6.4 Theorem (Hadamard).[15] *Suppose M is connected, simply connected, and complete with $K \leq 0$ everywhere on M. Then for every $p \in M$,*

(*) $\qquad\qquad\qquad \exp_p : T_p M \to M$

is a bijective diffeomorphism. (In other words, M is diffeomorphic to $\mathbb{R}^2$.) Moreover, there exists a unique minimal geodesic joining any two points, p and q, in M.

Remarks. 1. The theorem holds under the slightly weaker condition that there are no geodesic segments with conjugate points on M (cf. (6.8.4)).
2. The last statement sharpens the result (6.4.6) of Hopf and Rinow; we will prove this without using (6.4.6).
3. The fact that (*) is injective also follows from (6.3.3 (iv)).

PROOF. 1. The assumption that $K \leq 0$ allows us to use (6.5.6) and (6.5.2) to conclude that (*) is regular (maximal rank) and therefore is a *local* diffeomorphism. As in (6.2.2 (ii)), we can show that (*) is onto: Suppose $q \in M$. Consider a curve $b = b(s)$, $s_0 \leq s \leq s_1$, joining p to q. We may lift this curve to $T_p M$, via the inverse of $\exp_p$, to a curve $\tilde{b} = \tilde{b}(s)$, $s_0 \leq s \leq s_1$, which connects $0 \in T_p M$ to a point $\tilde{q} \in T_p M$. We have $\exp_p \tilde{b}(s) = b(s)$. This means that $q = \exp_p \tilde{q}$. *Note:* That $\exp_p$ is onto can also be deduced from (6.4.6).

2. We will now show that (*) is one-to-one: Suppose there exists $\tilde{q}_0, \tilde{q}_1 \in T_p M$ with $\exp_p \tilde{q}_0 = \exp_p \tilde{q}_1 = q \in M$. Let $\tilde{c}_i(t)$, $0 \leq t \leq 1$, be the line segments from 0 to $\tilde{q}_i$, $i = 0, 1$. Then $c_i(t) = \exp_p \tilde{c}_i(t)$, $i = 0, 1$, are geodesics from p to q. Since M is simply connected, c_0 is homotopic to c_1. This contradicts remark (3) to (6.6.3) unless $c_0 = c_1$. But $c_0 = c_1$ implies that $q_0 = q_1$.

3. Let $p \in M$. From the above discussion, it follows that $B_\rho(p)$ is a geodesic disk for all $\rho > 0$. Given $q \in M$, choose $\rho > d(p, q)$. Then $q \in B_\rho(p)$ and, by (5.3.4), there exists a unique geodesic from p to q.

6.7 Closed Geodesics and the Fundamental Group

In this section, M will always be assumed to be complete. A nonconstant geodesic $c = c(t)$, $0 \leq t \leq \omega$, is said to be *closed* of period ω if $\dot{c}(\omega) = \dot{c}(0)$. c is called *prime* if ω is the smallest positive number ω' such that $\dot{c}(\omega') = \dot{c}(0)$.

---

[15] Hadamard, J. Les surfaces à courbures opposées. *J. Math. Pures Appl.* (5) **4**, 27–73 (1898).

6.7 Closed Geodesics and the Fundamental Group

EXAMPLES. 1. Parameterized great circles are closed geodesics on the sphere $M = S^2$ whose period is 2π. Also, multiply covered great circles (of period $2\pi k$, $k = 1, 2, \ldots$) are closed geodesics.

2. Let M be the flat torus. M is the quotient of the Euclidean plane, $\mathbb{R}^2$, under the operation of $\mathbb{Z} \times \mathbb{Z}$ defined by

$$((m, n), (u, v)) \in \mathbb{Z} \times \mathbb{Z} \times \mathbb{R}^2 \mapsto (u + m, v + n) \in \mathbb{R}^2.$$

The geodesics of $\mathbb{R}^2$, the straight lines $u(t) = a_1 t + a_0$, $v(t) = b_1 t + b_0$ with $a_1^2 + b_1^2 \neq 0$, cover geodesics on M. The latter are closed geodesics if and only if a_1/b_1 or b_1/a_1 is rational.

We want to investigate how elements of the fundamental group of M and, more generally, certain fixed-point free isometries can give rise to closed geodesics.

Fix $p \in M$ and let $\alpha(t)$, $0 \leq t \leq 1$, be a continuous curve which begins and ends at p; $\alpha(0) = \alpha(1) = p$. Denote by $\Omega(p)$ the set of all such curves. If $\beta \in \Omega(p)$, we may consider the curve

$$\beta * \alpha(t) = \begin{cases} \alpha(2t), & \text{if } 0 \leq t \leq \tfrac{1}{2} \\ \beta(2(1-t)), & \text{if } \tfrac{1}{2} \leq t \leq 1, \end{cases}$$

that is, $\beta * \alpha$ is the curve α followed by β. Let $1 \in \Omega(p)$ be the constant curve. Denote by $[\alpha]$ the set of curves in $\Omega(p)$ which are homotopic to α via a homotopy which fixes p. The operation $*$ is associative up to homotopy. It is easy to check that if $[\beta] = [\beta']$ and $[\alpha] = [\alpha']$, then $[\beta * \alpha] = [\beta' * \alpha']$, $[1 * \alpha] = [\alpha]$; and if $\alpha^{-1}(t) = \alpha(1 - t)$, then $[\alpha * \alpha^{-1}] = [1]$. Therefore the operation $[\alpha] * [\beta] = [\alpha * \beta]$ is well defined and, with that operation, $\{[\alpha] | \alpha \in \Omega(p)\}$ is a group with identity $= [1]$ and $[\alpha]^{-1} = [\alpha^{-1}]$. This group is called the *fundamental group* of M at p, and is denoted by $\pi_1(p)$. See Massey, loc. cit., for more details.

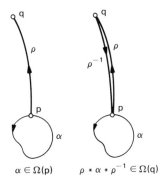

$\alpha \in \Omega(p)$ $\rho * \alpha * \rho^{-1} \in \Omega(q)$

Figure 6.8

If M is connected and q is another point of M, then $\pi_1(p)$ is isomorphic to $\pi_1(q)$. If ρ is a curve connecting p to q, then for $[\alpha] \in \pi_1(p)$, $[\rho * \alpha * \rho^{-1}] \in \pi_1(q)$,

6 The Global Geometry of Surfaces

and the map $[\alpha] \mapsto [\rho * \alpha * \rho^{-1}]$ is an isomorphism. In this case we write $\pi_1(M)$ for the fundamental group of M (at any point).

EXAMPLES. 1. $\pi_1(S^2) = \{[1]\}$, since every curve $\alpha \in \Omega(p)$, $p \in S^2$, is *contractible*, i.e., homotopic to the constant curve $c(t) \equiv p$.

2. $\pi_1(T^2) = \mathbb{Z} \times \mathbb{Z}$, where $\mathbb{Z}$ = integers. To see this, consider the standard torus in $\mathbb{R}^3$ (3.3.7). Since the fundamental group does *not* depend upon the choice of metric, but only upon the topology of the manifold, we are free to choose a convenient model. Let α denote the closed geodesic which forms the internal latitude line. Fix $p \in \alpha$ and let β be the longitude circle (a closed geodesic also) through p. If $\alpha^m = \alpha * \alpha * \ldots * \alpha$, $\alpha^{-m} = (\alpha^m)^{-1}$, etc., then it can be shown that every curve $\gamma \in \Omega(p)$ is homotopic to unique curve of the form $\alpha^m * \beta^n$ for some $(m, n) \in \mathbb{Z} \times \mathbb{Z}$ (proof left as an interesting exercise for the reader). This correspondence is a group isomorphism.

We shall assume the existence of a simply-connected covering surface upon which the fundamental group acts as deck transformations. (For a complete discussion, see Singer–Thorpe [A16].)

Let $\tilde{M}$ be the simply-connected covering surface of M and let

$$\mu: \tilde{M} \to M$$

be the covering projection—a local homeomorphism with the property that M possesses an atlas $(u_\alpha, M_\alpha)_{\alpha \in A}$ such that, for each $\alpha \in A$, $\mu^{-1}(M_\alpha)$ is a family $(\tilde{M}_{\alpha_i})_{i \in I}$ of open sets with $\mu | \tilde{M}_{\alpha_i}: \tilde{M}_{\alpha_i} \to M_\alpha$ being a diffeomorphism. We now define a differentiable atlas $(u_{\alpha_i}, \tilde{M}_{\alpha_i})_{(\alpha, i) \in A \times I}$ for $\tilde{M}$ with $u_{\alpha_i} = u_\alpha \circ (\mu | \tilde{M}_{\alpha_i})$. $u_{\beta_k} \circ u_{\alpha_i}^{-1} = u_\beta \circ u_\alpha^{-1}$ shows that this is indeed a differentiable atlas. The local representation of μ on U is $u_\alpha \circ \mu \circ u_{\alpha_i}^{-1} = \text{id}$. Thus $\mu: \tilde{M} \to M$ is differentiable. Moreover, we may define a Riemannian metric on M which will make μ a local isometry, i.e., for each $\tilde{p} \in \tilde{M}$, $d\mu_{\tilde{p}}: T_{\tilde{p}}\tilde{M} \to T_{\mu\tilde{p}}M$ shall be an isometry. This requires the following scalar product $g_{\tilde{p}}$ on $T_{\tilde{p}}\tilde{M}$:

$$g_{\tilde{p}}(\tilde{X}, \tilde{Y}) = g_{u_{\tilde{p}}}(d\mu_{\tilde{p}}\tilde{X}, d\mu_{\tilde{p}}\tilde{Y}) \quad \text{for all } \tilde{X}, \tilde{Y} \in T_{\tilde{p}}\tilde{M}.$$

Now let Γ denote the fundamental group M, considered as a group of deck transformations of $\tilde{M}$. In particular, if $\gamma \in \Gamma$, then $\mu \circ \gamma = \mu$. This implies that γ must be a local isometry of $\tilde{M}$.

The *conjugacy class* of γ is the set $\{\gamma' \gamma \gamma'^{-1} \mid \gamma' \in \Gamma\}$. For $\gamma = 1$, the conjugacy class is $\{1\}$.

We may now formulate and prove the main result of this section.

6.7.1 Theorem. *Let M be compact and let $\gamma \neq 1$ be an element of Γ, the fundamental group of M. Then there exists a γ-invariant geodesic $\tilde{c}$ in $\tilde{M}$, i.e., $\gamma\tilde{c}(t) = \tilde{c}(t + \omega)$ for all $t \in \mathbb{R}$. Here $|\dot{\tilde{c}}(t)| = 1$ and $\omega = \tilde{d}(\tilde{c}(0), \gamma\tilde{c}(0))$, where $\tilde{d}$ is the distance function on $\tilde{M}$. Under $\mu: \tilde{M} \to M$, $\tilde{c}$ projects onto a closed geodesic $c = \mu \circ \tilde{c}$ in M of period ω. The closed geodesic c is a representative of the conjugacy class of γ.*

PROOF. 1. On the universal covering surface $\tilde{M}$ of M, γ operates as a fixed-point free isometry: For suppose $\gamma(\tilde{p}) = \tilde{p}$. Then, if $(u_\alpha, \tilde{M}_\alpha)$ is a coordinate chart containing $\tilde{p}$, γ has the local expression $u_\alpha \circ \mu \circ \gamma \circ \mu^{-1} \circ u_\alpha^{-1} : U_\alpha \to U_\alpha$. But since $\mu \circ \gamma = \mu$, this map is equal to the identity. This means that $\gamma = $ identity near p. By the simple connectivity of $\tilde{M}$, $\gamma = $ id on $\tilde{M}$ which means $\gamma = 1$ is the neutral element of Γ. Contradiction.

2. Consider the function $f(\tilde{p}) = \tilde{d}(\tilde{p}, \gamma\tilde{p})$ on M. Since

$$\tilde{d}(\tilde{p}, \gamma\tilde{p}) \leq \tilde{d}(\tilde{p}, \tilde{q}) + \tilde{d}(\tilde{q}, \gamma\tilde{q}) + \tilde{d}(\gamma\tilde{q}, \gamma\tilde{p}) = \tilde{d}(\tilde{q}, \gamma\tilde{q}) + 2\tilde{d}(\tilde{p}, \tilde{q}),$$

f is a continuous function. Let $\{\tilde{p}_n\}$ be a sequence on M such that $\lim_n f(\tilde{p}_n) = \omega = \inf f$. Fix $\tilde{p}_0 \in \tilde{M}$, and let $d/2 = $ diameter of $M = $ maximum distance $d(p, q)$ between two points in M. Then, for every $n \geq 1$, there exists a $\gamma_n \in \Gamma$ such that $\tilde{d}(\gamma_n\tilde{p}_n, \tilde{p}_0) < d$. Therefore the sequence $\{\gamma_n\tilde{p}_n\}$ lies in the *compact* set $\{\tilde{p} \in \tilde{M} \mid d(\tilde{p}_0, \tilde{p}) \leq d\}$ (this set is bounded and $\tilde{M}$ is complete). Therefore $\{\gamma_n\tilde{p}_n\}$ has a limit point, say $\tilde{p}'$. For sufficiently large n, $\tilde{d}(\gamma\tilde{p}_n, \tilde{p}_n) = \tilde{d}((\gamma_n\gamma\gamma_n^{-1})\gamma_n\tilde{p}_n, \gamma_n\tilde{p}_n)$ is near ω and $\gamma_n\tilde{p}_n$ is near $\tilde{p}'$. Thus for large n, $\gamma_n\gamma\gamma_n^{-1}\tilde{p}'$ is near to $\tilde{p}'$. But within a fixed distance of $\tilde{p}'$ there can only be a finite number of *different* points $\gamma_n\gamma\gamma_n^{-1}\tilde{p}'$. Therefore there exists a $\gamma_0 \in \Gamma$ such that for an infinite number of n, $\gamma_n\gamma\gamma_n^{-1}\tilde{p}' = \gamma_0\gamma\gamma_0^{-1}\tilde{p}'$. But this means that

$$d(\gamma_0\gamma\gamma_0^{-1}\tilde{p}', \tilde{p}') = \omega.$$

Define $\gamma_0^{-1}\tilde{p}' = \tilde{p}$.

3. Let $\tilde{c}(t)$, $t \in \mathbb{R}$, be a geodesic with $\tilde{c}(0) = \tilde{p}$, $\tilde{c}(\omega) = \gamma\tilde{p}$. Since $\tilde{M}$ is geodesically complete, such a curve exists by (6.4.6). We will prove that $\gamma\tilde{c}(t) = \tilde{c}(t + \omega)$ for all t. By definition this is true for $t = 0$. It will also hold for $t \in [0, \omega]$ unless the geodesics $\tilde{c}(t + \omega)$, $t \geq 0$, and $\tilde{c}(t)$, $t \geq 0$, have different tangent vectors at their common initial point $\tilde{c}(\omega)$. But if this were so, then we would have

$$\tilde{d}(\tilde{c}(t), \gamma\tilde{c}(t)) < \tilde{d}(\tilde{c}(t), \tilde{c}(\omega)) + \tilde{d}(\gamma\tilde{c}(0), \gamma\tilde{c}(t)) = \omega,$$

which contradicts the definition of ω. Thus $\gamma\tilde{c}(t) = \tilde{c}(t + \omega)$ on $[0, \omega]$ and hence for all $t \in \mathbb{R}$.

The image $c(t) = \mu\tilde{c}(t)$ of $\tilde{c}$ under μ is therefore a closed geodesic: $c(t + \omega) = c(t)$. □

Remark. The geodesic $\tilde{c}(t)$ need not be unique. For example, consider the flat torus whose universal covering is the Euclidean plane (see example 2 above). If $\tilde{c}(t)$ is a line in the plane invariant under $\gamma \neq 1$, then any integral translation of $\tilde{c}(t)$ is also γ-invariant. A similar situation holds true for the projective plane, covered by S^2 (see (5.5.3, 2)).

A further existence theorem for closed geodesics on compact surfaces M is the following.

6.7.2 Theorem. *Suppose $\gamma: M \to M$ is an isometry of M which has no fixed points. Then there exists a γ-invariant geodesic $c = c(t), t \in \mathbb{R}$, i.e., $\gamma \circ c(t) = c(t + \omega)$ for all t. If γ is of finite order, i.e., if there exists an $n > 1$ such that $\gamma^n = 1$, then c is closed with period $n\omega$.*

PROOF. Consider the function $f(p) = d(p, \gamma p)$. As in the proof of (6.7.1), we can easily show that f is continuous. Since M is compact, f assumes a minimum value, say ω, at some point p. Since γ is fixed-point free, $\omega = d(p, \gamma p) > 0$. Let $c(t)$ be a unit-speed geodesic with $c(0) = p$ and $c(\omega) = \gamma p$. As in the proof of (6.7.1), we can show that $\gamma \circ c(t) = c(t + \omega)$. If $\gamma^n = \text{id}$, then it follows immediately that $c(t)$, $0 \le t \le n\omega$, is closed. □

EXAMPLES. 1. The flat torus M with universal covering $\tilde{M} = \mathbb{R}^2$. (Example 1 above.) Every translation $\tilde{\tau}: \mathbb{R}^2 \to \mathbb{R}^2$ induces an isometry τ on M. The isometry τ is the identity if and only if $\tilde{\tau}(0, 0) \in \mathbb{Z} \times \mathbb{Z}$. Suppose $\tilde{\tau}$ does not generate the identity. Then there exists a τ-invariant closed geodesic if and only if $\tilde{\tau}$ satisfies

$$\tilde{\tau}(0, 0) \in \mathbb{Q} \times \mathbb{Q} - \mathbb{Z} \times \mathbb{Z},$$

where $\mathbb{Q} = $ field of rationals.

2. The sphere S^2 (see 5.7). The antipodal map is an isometry of order two. The closed geodesics, whose existence is proved in the above theorem, are the great circles.

The results of this section may be generalized to Riemannian manifolds. The interested reader is referred to Kobayashi, S. *Differential Geometry and Transformation Groups* [B14], Chapter 3. A few of these theorems can be found in section 6.8 below.

To conclude, we prove a theorem of Preissmann which makes explicit the consequences of Theorem (6.7.1) for compact surfaces with $K \le 0$.

6.7.3 Theorem (Preissmann).[16] *Let M be a compact surface with $K \le 0$. Then:*
  i) *Γ is infinite.*
  ii) *Every element $\gamma \ne 1$ of Γ has infinite order.*
  iii) *For each $\gamma \ne 1$ in Γ there exists a γ-invariant unit-speed geodesic in $\tilde{M}$. If $K < 0$, this geodesic is unique up to choice of initial point.*
  iv) *If $K < 0$, every abelian subgroup of Γ is an infinite cyclic group.*

Remark. If $K \le 0$, the conclusions of statements (iii) and (iv) *need not be true*. They fail, for example, on the flat torus, Example 1 above.

PROOF. 1. By (6.6.4), the universal covering $\tilde{M}$ of M is diffeomorphic to $\mathbb{R}^2$ and therefore is noncompact. It follows that Γ is infinite.

2. Let $\gamma \in \Gamma$, $\gamma \ne 1$. By (6.7.1), there exists a γ-invariant geodesic $\tilde{c}: \gamma\tilde{c} = \tilde{c}(t + \omega)$ for all t, where $\omega = d(\tilde{c}(0), \tilde{c}(\omega))$. By (6.6.4), we may conclude that for all positive integers n, $\gamma^n \tilde{c}(0) = \tilde{c}(n\omega) \ne \tilde{c}(0)$. Therefore γ has infinite order.

[16] Preissmann, A. Quelques propriétés globales des espaces de Riemann. *Comment. Math. Helv.* **15**, 175–216 (1943). See also Cartan [B4], note III. The proof given there is unfortunately not completely correct, and the claimed existence of two geodesics in a given homotopy class is false if $K < 0$.

3. Suppose $K < 0$. To prove (iii), it suffices to show that if $\tilde{c}(t)$ and $\tilde{c}'(t)$, $t \in \mathbb{R}$, are γ-invariant geodesics in $\tilde{M}$, i.e.,

$$\gamma\tilde{c}(t) = \tilde{c}(t + \omega); \qquad \gamma\tilde{c}'(t) = \tilde{c}(t + \omega'),$$

then $\tilde{c}'(t) = \tilde{c}(t + t_0)$ for some fixed t_0.

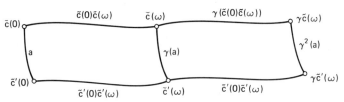

Figure 6.9 Geodesic quadrilaterals

To see this, consider the geodesic quadrilateral with vertices $\tilde{c}(0)$, $\tilde{c}(\omega)$, $\tilde{c}'(\omega)$, and $\tilde{c}'(0)$. By (6.4), this figure is uniquely defined. If this quadrilateral were nondegenerate, the sum of the angles at the vertices would be 2π. This is because γ is an isometry which maps the edge $\tilde{c}(0)\tilde{c}'(0)$ into the edge $\tilde{c}(\omega)\tilde{c}'(\omega')$ and maps the geodesics $\tilde{c}$ and $\tilde{c}'$ into themselves. But by Corollary (6.3.3 (ii)) to the Gauss–Bonnet theorem, the sum of the interior angles of a geodesic quadrilateral must be $< 2\pi$ when $K < 0$. Therefore the quadrilateral is degenerate which means there must exist a $t_0 \in \mathbb{R}$ with $\tilde{c}'(t) = \tilde{c}(t + t_0)$ for all t.

4. Suppose γ and γ' are nontrivial commuting elements in Γ. Let $\tilde{c}(t)$, $t \in \mathbb{R}$, and $\tilde{c}'(t)$, $t \in \mathbb{R}$, be the corresponding invariant geodesics in $\tilde{M}$. By (iii), they are unique up to choice of initial points. Since $\gamma\gamma' = \gamma'\gamma$, $\gamma\gamma'\tilde{c} = \gamma'\gamma\tilde{c} = \gamma'\tilde{c}$. In other words, $\gamma'\tilde{c}$ is γ-invariant. By (iii), we may conclude that $\gamma'\tilde{c} = \tilde{c}$ up to choice of initial point. But this means that $\tilde{c}$ is γ'-invariant, so, by (iii), $\tilde{c} = \tilde{c}'$ up to choice of initial point. We reparameterize so that $\tilde{c}(t) = \tilde{c}'(t)$ for all t. Let $B_\rho(\tilde{c}(t))$ be a geodesic disk centered at $\tilde{c}(t)$ for any fixed t. If $\rho > 0$ is sufficiently small, then for all integers k and l, either $\gamma^k\gamma'^l\tilde{c}(t) = \tilde{c}(t + k\omega + l\omega')$ is equal to $\tilde{c}(t)$ or $\gamma^k\gamma'^l\tilde{c}(t)$ lies *outside* $B_\rho(\tilde{c}(t))$. Therefore there must be some $\omega_0 > 0$ so that $\omega = m\omega_0$ and $\omega' = m'\omega_0$ for some integers m, m'. Thus $\gamma^k\gamma'^l\tilde{c}(t) = \tilde{c}(t + (km + lm')\omega_0)$ for all k, l integers. Therefore there must be an element $\gamma_0 \in \Gamma$ which generates a cyclic group (infinite by (ii)) containing γ and γ'. The element $\gamma_0 \in \Gamma$ is determined by the equation $\gamma_0\tilde{c}(t) = \tilde{c}(t + \omega_0)$. □

6.8 Exercises and Some Further Results

6.8.1 Recall that an ovaloid is a compact surface in $\mathbb{R}^3$ with $K > 0$ ((6.2)). As a surface with Riemannian metric, it must be diffeomorphic to S^2 by (6.3.5). A natural question to ask is: Given a surface M diffeomorphic to S^2 and endowed with a metric for which $K > 0$, does there exist an ovaloid in $\mathbb{R}^3$ which is isometric to M? The answer is yes. This existence theorem was partially proved by H. Weyl. A complete proof in the real analytic case was given by H. Lewy. The theorem for differentiable M was proved independently by Alexandrov, working with Pogorelov, and by Nirenberg. Their respective proofs are quite different in method.[17]

6.8.2 The second part of the Sturm comparison theorem (6.5.5) and its application (6.5.6) has the following geometric interpretation.

Suppose M and M^* are surfaces with Riemannian metric whose curvature functions K and K^* satisfy $\max K \leq \min K^*$, which we will write for short as $K \leq K^*$. Suppose c and c^* are two geodesic arcs, parameterized by arc length, on M and M^*, respectively, whose lengths are both equal to a. Assume that $a \leq \pi/\sqrt{\max K^*}$. This insures that both segments are free of conjugate points in their interior. Suppose further that $Y(t)$ and $Y^*(t)$ are Jacobi fields along $c(t)$ and $c^*(t)$, respectively, with $Y(0) = Y^*(0)$ and $|\nabla Y(0)| = |\nabla Y^*(0)|$. Then $|Y(t)| \geq |Y^*(t)|$.

This is the infinitesimal version of the Alexandrov comparison theorem for geodesic triangles.[18] We will state one special case of the theorem here. The proof, which involves introducing polar coordinates and integration of the inequality of the above theorem, is left as an exercise.

Under the conditions and assumptions above, consider two geodesic arcs emanating from a point $p \in M$. Denote these geodesics by c and c' and suppose that they have end points q and q', respectively. Let the length of c equal that of c' and denote their common length by a which we will assume is equal to or less than $\pi/\sqrt{\max K^*}$. Let c^* and $c^{*\prime}$ be two geodesic segments in M^*, emanating from a point $p^* \in M^*$, whose lengths are also equal to a and whose end points are q^* and $q^{*\prime}$, respectively. Suppose the angle at p between c and c' is equal to the angle at p^* between c^* and $c^{*\prime}$. Then if this angle is sufficiently small, $d(q,q') \geq d(q^*,q^{*\prime})$. (See Figure 6.10).

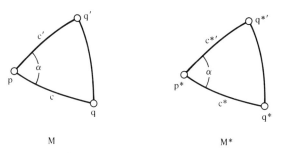

Figure 6.10   Geodesic triangles

---

[17] Weyl, H. *Über die Bestimmung einer geschlossenen konvexen Fläche durch ihr Linienelement*. Vierteljahrsschrift Naturforsch. Gesellschaft Zürich, 1916, 40–72. Lewy, H. On the existence of a closed surface realizing a given Riemannian metric. *Proc. Nat. Acad. Sci. U.S.A.* **24**, 104–106 (1938). Alexandrov, A. D. [B1]. Pogorelov, A. V. *Deformation of convex surfaces*. Gosudarstv. Izdat. Tehn-Teor. Lit., Moscow-Leningrad (1951) (Russian). English review: **MR 12**, 400. German translation: Berlin, Akademie-Verlag, 1955. Nirenberg, L. The Weyl and Minkowski problems in differential geometry in the large. *Comm. Pure Appl. Math.* **6**, 337–394 (1953).

[18] Alexandrov, A. D. [B1].

6.8.3 Suppose M is a surface with Riemannian metric which is diffeomorphic to S^2 and on which $K \leq K_1$. The example of the hourglass surface in (6.6.1) illustrates that it is not possible to estimate the injectivity radius solely on the basis of an upper bound on the curvature alone.

However, if the curvature also satisfies $0 < K_0 \leq K$, we can say something. First of all, M can be realized as an ovaloid in $\mathbb{R}^3$ (see 6.8.1). A result due to Pogorelov states that for all $p \in M$ the injectivity radius $\rho_m(p)$ is at least $\pi/\sqrt{K_1}$.[19] On the other hand, we know from (6.5.8) that the *most* it can be is $\pi/\sqrt{K_0}$. Therefore $\pi/\sqrt{K_1} \leq \rho_m(p) \leq \pi/\sqrt{K_0}$ for all $p \in M$. This implies that the intrinsic diameter $d(M) = \sup_{q, p \in M} d(q, p)$ also satisfies $\pi/\sqrt{K_1} \leq d(M) \leq \pi/\sqrt{K_0}$.

The example of the sphere of constant curvature K_0 (resp. K_1) shows that these bounds are best possible. For a sphere of curvature K', $\rho = \rho_m(p) = \pi/\sqrt{K'}$.

The theorem of Pogorelov is equivalent to the fact that on an ovaloid M with $K_0 \leq K \leq K_1$, a closed geodesic must have length at least $2\pi/\sqrt{K_1}$. Moreover, a closed geodesic on M which has no self-intersection can have length at most $2\pi/\sqrt{K_0}$. These estimates are sharp, as the example of a sphere of constant curvature shows.

6.8.4 *Show:* Suppose M is a complete, simply-connected surface with a Riemannian metric. If for some $p \in M$ all geodesic rays emanating from p are free of conjugate points, then $\exp_p \colon T_pM \to M$ is a diffeomorphism, and therefore the injectivity radius $\rho_m(p) = \infty$. (*Hint:* Use the lifting technique of (6.6.3).)

6.8.5 In (6.7), we proved the existence of closed geodesics on compact surfaces which were not simply connected. It turns out that closed geodesics always exist, even on compact simply-connected surfaces—i.e., surfaces which are diffeomorphic to S^2. The proof of this fact requires techniques beyond those developed in this book.

In fact, if M is diffeomorphic to S^2 there must exist at least three different simple closed geodesics. This result is due to Lusternik and Schnirelmann. Moreover, there exists such a surface with exactly three simple closed geodesics and no more.

Consider an ellipsoid with three different axes. If the ratios of the lengths of the axes are sufficiently close to 1, then the only simple closed geodesics are the ellipses which occur as the intersection of the ellipsoid with the coordinate planes.[20]

---

[19] Pogorelov, A. V. A theorem regarding geodesics on closed convex surfaces. *Math. Sb.* N.S. (**18**), (60), 181–183 (1946) (Russian with English summary). English review: MR **8**, 16. The proof is not quite complete. For a different proof, see Klingenberg, W. Neue Ergebnisse über konvexe Flächen. *Comm. Math. Helv.* **34**, 17–36 (1960).

[20] Lusternik, L., and Schnirelmann, L. Sur le problem de trois géodesiques fermées sur surfaces de genus 0. *C.R. Acad. Sci. Paris* **189**, 269–271 (1929). An excellent presentation of this and other results may be found in Lusternik, L. The topology of function spaces and the calculus of variations in the large. *Trudy Math. Inst. Steklov* **19** (1947) (Russian)—translated into English in Translations of Math. Monographs, Vol. 16, A.M.S., Providence, R.I., 1966. See also the forthcoming book Klingenberg, W., *Lectures on Closed*

6   The Global Geometry of Surfaces

6.8.6   Suppose M is compact and $K < 0$ on M. It follows from (6.7.1) and (6.7.3) that every nontrivial conjugacy class in the fundamental group Γ corresponds to exactly one closed geodesic (up to parameterization). It can be deduced from the structure of the fundamental group of such a surface (which must have negative Euler characteristic) that there must be an infinite number of *different* unparameterized prime closed geodesics on M. This means that we count only those closed geodesics which are not a covering of some other closed geodesic. For a discussion of the structure of Γ for a surface M with $K < 0$, see Seifert-Threlfall, *Lehrbuch der Topologie*, Chelsea, New York, or Kobayashi [B14].

This result has been strengthened by E. Hopf, who proved that the subset Per $TM = \{X \in TM \mid \exp tX, t \in \mathbb{R}$ is a closed geodesic$\}$ is *dense* in TM.[21]

6.8.7   Let M be a compact surface with Riemannian metric. A pair of distinct points (p, p') in M is called a "Wiedersehen" pair if each geodesic emanating from p passes through p' and conversely. For example, the north and south of a compact surface of revolution is a "Wiedersehen" pair.

Prove: On the ellipsoid (3.7.3), both pairs of diametrically opposite umbilics are "Wiedersehen" pairs. See (3.9.5.).

An oriented surface M is called a "Wiedersehensfläche" if every $p \in M$ belongs to a "Wiedersehen" pair ("fläche" = surface in German). This name is due to Blaschke, who observed that such a surface must be homeomorphic to S^2. It was a long-standing open problem as to whether or not a "Wiedersehensfläche" was necessarily isometric to a sphere with constant curvature. The question was resolved, affirmatively, in 1963 by L. Green in a paper with the punning title "Auf Wiedersehensflächen."[22] (It is in English.)

6.8.8   There exist compact surfaces on which *all* geodesics are closed and have the same length but which are not isometric to a sphere of constant curvature. The first examples of such surfaces were constructed by Zoll, who used an idea due to Darboux. The surfaces are called Zoll surfaces. Recently, Riemannian manifolds with the same property have been investigated by Weinstein, Berger, and others.[23]

Geodesics. Springer-Verlag, Berlin-Heidelberg-New York, 1978. Jacobi had already investigated the behavior of geodesics on ellipsoids in his Vorlesungen über Mechanik, Winter-Semester 1842/43, Königsberg. See Darboux [A7], Volume III, Book VI, Chapter 1.

[21] Hopf, E. Statistik der geodätischen Linien in Mannigfaltigkeiten negativer Krümmung. *Ber. Verh. Sächs. Akad. Wiss. Leipzig* **91**, 261–304 (1939). For more recent developments, see Anosov, D. V. Geodesic flows on closed Riemannian manifolds with negative curvature. *Trudy Mat. Inst. Steklov* **90** (1967) (Russian)—English translation: *Proc. Steklov Inst. Math.* **90** (1967), A.M.S., Providence, R.I., 1969.

[22] Green, L. Auf Wiedersehensflächen. *Ann. of Math.* **78**, 289–299 (1963).

[23] See Darboux [A7], Part III, Book 6, Chapter 1. Zoll, O. Über Flächen mit Scharen geschlossener geodätischer Linien. *Math. Ann.* **57**, 108–133 (1903). See also Berger, M. *Lectures on Geodesics in Riemannian Geometry.* Tata Institute of Fundamental Research, Bombay, 1965, and Besse, A., *Manifolds all of whose Geodesics are Closed.* Springer-Verlag, Berlin-Heidelberg-New York, 1977.

6.8.9 Suppose M is a surface which is homeomorphic to a torus. By the Gauss-Bonnet theorem for compact surfaces, (6.3.5), any metric on M must satisfy $\int_M K \, dM = 0$. Thus, if there is a $p \in M$ with $K(p) > 0$, there must also be a $p' \in M$ with $K(p') < 0$.

The flat torus satisfies $K \equiv 0$ (see the second example in (6.7)). By (6.5.6 (i)), the flat torus is free of conjugate points. A converse of this result has been proved by E. Hopf: Suppose M is homeomorphic to a torus and let M have a Riemannian metric, g, in which no geodesic has a conjugate point. Then this metric satisfies $K \equiv 0$, i.e., (M, g) is the flat torus.[24]

6.8.10 Suppose M is a compact Riemannian surface. For any $p \in M$, the *cut locus*, $C(p)$, of p is defined as follows:

Associated to each tangent vector $X \in S_p^1 M \subset T_p M$ there exists a well-defined extended real number $t(X) > 0$ for which

a) the unique geodesic $c_X(t) = \exp_p tX$ with initial condition X is length-measuring on $[0, t(X)]$;

b) for every $t > t(X)$, $d(p, c_X(t)) < t$.

The map $c_R(t) \mapsto t(X)X \in T_p M$ is continuous. The image of this map is a non-self-intersecting closed curve in $T_p M$. The image of this curve under $\exp_p$ is $C(p) = \{c_X(t(X)) | X \in S_p^1 M\}$.

The complement of $C(p)$ in M may be contracted *radially* onto p: Each such point q is of the form $\exp_p(t_0 X_0)$, $t_0 < t(X_0)$, and $c_{X_0}(t)$, $0 \leq t \leq t_0$ is a minimal geodesic from p to q. Contract by sliding q along $c_{X_0}(t)$ to p. Thus $M \setminus C(p)$ is homeomorphic to the 2-cell $B_1(0) = \{X \in T_p M; |X| < 1\}$. We may consider M as $B_1(0)$ modulo the following identification of the boundary points $\partial B_1(0) = S_p^1 M$: Set $X \sim X'$ if $\exp_p(t(X)X) = \exp_p(t(X')X') \in C(p)$. Note: If $q \in C(p)$, there need not exist more than one minimizing geodesic from p to q.

The topology of M is completely determined by the topological structure of $C(p)$. For example, if $M = S^2$, a sphere of constant curvature, $C(p)$ is equal to the antipodal point of p. If M is the flat torus (see (6.7)), $C(p)$ consists of two circles which cross at one point. The same is true for the standard embedding of the torus in $\mathbb{R}$ (see 3.3.7).

The cut-locus was first investigated by Poincaré (he called it "ligne de partage").[25] Myers and others continued the study and clarified the concepts. For a detailed discussion of the cut-locus, see the article by Kobayashi in [A6]. For a description of classical as well as recent results we refer the reader to an article of H. Karcher.[26]

More recently, it has been shown by Buchner *et al.* that the cut-locus of a Riemannian manifold is *stable* in the following sense: Let M be a

---

[24] Hopf, E. Closed surfaces without conjugate points. *Proc. Nat. Acad. Sci. U.S.A.* **34**, 47–51 (1948).

[25] Poincaré, H. Sur les lignes géodesiques des surfaces convexes. *Trans. Amer. Math. Soc.* **6**, 237–274 (1905).

[26] Karcher, H. Schnittort und konvexe Mengen in vollstandigen riemannschen Mannigfaltigkeiten. *Math. Ann.* **177**, 105–121 (1968), and also Anwendungen der Alexandrowschen Winkelvergleichsatze. *Manuscripta Math.* **2**, 77–102 (1970).

manifold with a Riemannian metric g. Consider $p \in M$ and $C(p) \subset M$. Let $\tilde{g}$ be another Riemannian metric which is close to g in some natural sense, and let $\tilde{C}(p)$ be the cut-locus of p in the $\tilde{g}$ metric. Then for almost all $p \in M$ and a large *generic* class of metrics on M, there exists a homeomorphism $\phi: M \to M$ such that $\phi | C(p) \to \tilde{C}(p)$ is also a homeomorphism.

However, it is possible to construct cut loci which are not *triangulable*, i.e., cannot be decomposed into polygons. This has been done for surfaces by Gluck and Singer.[27]

6.8.11 Open surfaces in the large. A detailed study of complete *open* (i.e., non-compact) surfaces was initiated by Cohn-Vossen. We mention only the following result. Suppose M is a complete open surface on which the Gauss curvature is everywhere positive. Then M is diffeomorphic to the plane and $\int_M K \, dM \leq 2\pi$. Moreover, there are *no* closed geodesics on M and every geodesic has at most one self-intersection from which it runs off to infinity in both directions (it leaves every compact subset of M). Any two complete geodesics must intersect and through each $p \in M$ there passes at least one complete geodesic without self-intersection.[28] An example (really *the* example!) of such a surface on which all these properties may be verified directly is the paraboloid of revolution in $\mathbb{R}^3$.

6.8.12 (i) Let $c(t)$ be a unit-speed curve in $\mathbb{R}^n$ with the property that $|c(t)|^2$ has a local maximum at t_0. Let $p_0 = c(t_0)$ and $\rho^2 = |p_0|^2$. Show: $\kappa(t_0) \geq 1/\rho$, where $\kappa(t_0) = |\ddot{c}(t_0)|$ (which is equal to the first curvature of $c(t)$ at t_0 if it is defined).

(ii) Let M be a surface in $\mathbb{R}^n$ with $M \subset \{x \in \mathbb{R}^n | |x| \leq \rho\}$. Show: If $p_0 \in M$ satisfies $|p_0| = \rho$, then any curve $c(t)$ on M with $c(0) = p_0$ must have normal curvature with absolute value not less than $1/\rho$ at $t = 0$. Moreover, the sign of the normal curvature at $t = 0$ will be the same for any such curve.

(iii) Let M and $p_0 \in M$ be as in (ii). Show: $K(p) \geq 1/\rho^2$.

(iv) The map $\det: \mathrm{GL}(n, \mathbb{R}) \to \mathbb{R}$ is differentiable since the determinant of a matrix is a polynomial in the entries of the matrix. Show: Every value of $\det: \mathrm{GL}(n, \mathbb{R}) \to \mathbb{R}$ is a regular value. (*Hint:* Consider $A \in \mathrm{GL}(n, \mathbb{R})$ as $(A^1, \ldots, A^n)$ where A^i is the ith column of A. Then $\det(A^1, \ldots, tA^i, \ldots, A^n) = t \det A$. Use this fact to find a tangent vector X to $\mathrm{GL}(n, \mathbb{R})$ at A satisfying $d(\det)_A(X) \neq 0$. See (6.1.5, 3).)

---

[27] Singer, D., and Gluck, H. The existence of non-triangulable cut loci. *Bull. Amer. Math. Soc.* **82**, 4, July 1976, pp. 599–602. Buchner, M. Thesis, Harvard University, 1974.

[28] These last results may be generalized to complete, open Riemannian manifolds of positive curvature. See Gromoll, D., and Meyer, W. On complete open manifolds of positive curvature. *Ann. of Math.* **90**, 75–90 (1969).

References

A. Texts and surveys of differential geometry

[A1] Auslander, L. *Differential Geometry.* New York, N.Y.: Harper and Row, 1967.

[A2] Blaschke, W. Vorlesungen über Differentialgeometrie. Band 1. *Elementare Differentialgeometrie,* 4. Aufl. Berlin: Springer, 1945.

[A3] Blaschke, W., and Reichardt, H. *Einführung in die Differentialgeometrie,* 2. Aufl. Berlin–Göttingen–Heidelberg. Springer, 1960.

[A4] Blaschke, W., Kreis und Kugel. Leipzig 1916. Second edition: de Gruyter and Co., Berlin, 1956. 5th Edition (with K. Leichtweiss) Springer, 1973.

[A5] Chern, S. S. *Differential Geometry.* Lecture Notes, Dept. of Mathematics, University of Chicago, 1954.

[A6] Chern, S. S. *Studies in Global Geometry and Analysis.* The Mathematical Association of America, Englewood Cliffs, N.J.: Prentice-Hall, 1967.

[A7] Darboux, G. *Leçons sur la théorie générale des surfaces,* Volumes I to IV. Paris: Gauthier-Villars, 1887–1896. Reprint of 3rd edition, New York: Chelsea, 1972.

[A8] do Carmo, M. *Differential Geometry of Curves and Surfaces.* Englewood Cliffs, N.J.: Prentice-Hall, 1976.

[A9] Haack, W. *Elementare Differentialgeometrie.* Basel and Stuttgart: Birkhäuser, 1955.

[A10] Hilbert, D., and Cohn-Vossen, S. *Anschauliche Geometrie.* Berlin: Springer, 1932.

[A11] Hopf, H. *Selected Topics in Differential Geometry in the Large.* Notes by T. Klotz, New York, N.Y.: Institute of Mathematical Sciences, New York University, 1955.

[A12] Hopf, H. *Lectures on Differential Geometry in the Large.* Notes by J. W. Gray, Stanford, Calif.: Applied Mathematics and Statistics Laboratory, Stanford University, 1955.

References

[A13] Laugwitz, D. *Differential and Riemannian Geometry.* Academic Press, 1965.

[A14] Nirenberg, L. *Seminar on Differential Geometry in the Large.* New York, N.Y.: Institute for Mathematical Sciences, New York University, 1956.

[A15] O'Neill, B. *Elementary Differential Geometry.* New York, N.Y.: Academic Press, 1966.

[A16] Singer, I. M., and Thorpe, J. A. *Lecture Notes on Elementary Topology and Geometry.* Glenview, Ill.: Scott, Foresman and Co., 1967; Springer-Verlag, 1976.

[A17] Spivak, M. *A Comprehensive Introduction to Differential Geometry,* Vol. I–V. Boston, Mass.: Publish or Perish, 1972–1975.

[A18] Strubecker, K. *Differentialgeometrie,* 3 Bd., 2. Aufl. Berlin: De Gruyter und Co., 1964, 1969.

[A19] Struik, D. J. *Lectures on Classical Differential Geometry,* 2nd Edition. Reading, Mass.: Addison-Wesley, 1961.

[A20] Willmore, T. *An Introduction to Differential Geometry.* Oxford: Clarendon Press, 1959.

B. *Some more advanced texts and monographs*

[B1] Alexandrov, A. D. *The Intrinsic Geometry of Convex Surfaces.* Moscow-Leningrad: Gosudarstv. Izdat Tehn-Teor. Lit., (1948). (Russian) German translation: *Die Innere Geometrie konvexer Flächen.* English Review (**MR 10,** 619 and **17,** 74).

[B2] Bishop, R., and Crittenden, R. *Geometry of Manifolds.* New York, N.Y.: Academic Press, 1964.

[B3] Berger, M., Gauduchon, P., and Mazet, E. *Le spectre d'une variété riemannienne.* Springer Lecture Notes 194, Berlin-Heidelberg-New York: Springer 1971.

[B4] Cartan, E. *Leçons sur la géometrie de Riemann,* 2nd Edition Paris: Gauthier-Villars, 1946.

[B5] Chern, S. S. *Topics in Differential Geometry.* Princeton, N.J.: Institute for Advanced Study, 1951.

[B6] Chern, S. S. *Differentiable Manifolds.* Lecture Notes, Dept. of Mathematics, University of Chicago, 1959.

[B7] Duschek, A., and Mayer, W. *Lehrbuch der Differentialgeometrie,* Band II: *Riemannsche Geometrie von W. Mayer.* Leipzig and Berlin: Teubner, 1930.

[B8] Flanders, H. *Differential Forms.* New York, N.Y.: Academic Press, 1963.

[B9] Gromoll, D., Klingenberg, W., and Meyer, W. *Riemannsche Geometrie im Großen.* Berlin–Heidelberg–New York: Springer, 1968.

[B10] Helgason, S. *Differential Geometry and Symmetric Spaces.* New York, N.Y.: Academic Press, 1962.

[B11] Hermann, R. *Differential Geometry and the Calculus of Variations.* New York, N.Y.: Academic Press, 1968.

References

[B12] Hicks, N. J. *Notes on Differential Geometry.* Princeton, N.J.: Van Nostrand, 1965.

[B13] Kobayashi, S., and Nomizu, K. *Foundations of Differential Geometry*, 2 Volumes. New York, N.Y.: Interscience, 1963 and 1969.

[B14] Kobayashi, S. *Differential Geometry and Transformation Groups.* Berlin-Heidelberg-New York: Springer 1972.

[B15] Lang, S. *Differential Manifolds.* Reading, Mass.: Addison-Wesley, 1972.

[B16] de Rham, G. *Variétés différentiables.* Paris: Hermann, 1955.

[B17] Sternberg, S. *Lectures on Differential Geometry.* Englewood Cliffs, N.J.: Prentice-Hall, 1964.

[B18] Vranceanu, G. *Leçons de Geometrie Differentielle*, I–III. Bukarest: Editions de L'Academie de la Republique Populaire Roumaine, 1957–1964. German translation: Akademie-Verlag, Berlin, 1961.

[B19] Warner, F. *Foundations of Differentiable Manifolds.* Glenview, Ill.: Scott Foresman Co., 1972.

Index

adjoint transpose of matrix 132
Alexandrov, A. D. 162, 168
Alexandrov comparison theorem 162
angle between vectors 76
Anosov, D. V. 164
antipodal map 111
area element 70, 115
asymptotic directions 53, 54
asymptotic lines 54
atlas 105, 124
average curvature 31

Barner, M. 30
base point of tangent vector 96
Berger, M. 120, 164, 168
Bernstein's theorem 70, 71
Besse, A. 164
bilinear forms 35, 111
Bishop, R. 149, 153, 168
Blaschke, W. 29, 30, 129, 135, 164, 167
Bonneson, T. 31
Bonnet, O. 139, 152
Buchner, M. 165, 166

canal surface 67, 68
canonical projection 5
catenoid 70
Cauchy–Riemann equations 121
caustic surface 67, 68
change of coordinates for curves 8
change of coordinates for surfaces 34
charts on a manifold 105, 123

Chern, S. S. 29, 31, 68, 70, 130, 167, 168
Christoffel symbols 61, 74, 91
circle, geodesic 99, 100
circle of hyperbolic geometry 30
circle in the plane 17
Clairaut's theorem 87, 88, 122
closed curve 21
closed geodesics 156
closed set 2
Codazzi–Mainardi equation 62
Cohn-Vossen, S. 135, 166, 167
comparison theorems 151, 162
completeness 146
cone 57
confocal surfaces 55, 56
conformal coordinates 120
conformal surfaces 93
congruence 2
conjugacy class 158
conjugate harmonic function 120
conjugate point 148
constant Gauss curvature 66, 93
constant curvature surfaces 83–88, 137
constant width curve 30
continuously differentiable map 4
continuous map 2
contractible space 158
convex curves 27–29
convex surfaces 129
coordinate charts 105, 123
coordinate invariance of covariant differentiation 91

171

Index

coordinate invariance of curvature tensor 92
coordinate invariance of energy 91
coordinate invariance of Frenet frame 91
coordinate invariance of Gauss curvature 92
coordinates, asymptotic 53
coordinates, change, 8, 34, 74, 90, 113
coordinates, conformal 120
coordinates, Fermi 81
coordinates, geodesic orthogonal 80–82
coordinates, geodesic polar 99–101
coordinates, isothermal 120
coordinates, orthogonal 77
coordinates, principal curvature 53
coordinates, Riemann normal 99
coordinate systems 105, 123
coordinate vector field 34
corners of a curve 27–29
Courant, R. 71
covariant differential 76, 91
covariant derivative 74, 75, 91
covering projection 158
covering space 158
Crittenden, R. 149, 153, 168
cross product 4
curvature, average 31
curvature, constant 83–88, 137
curvature, constant Gauss 66, 93
curvature, Gauss 30, 47, 56, 64, 91
curvature, geodesic 30, 47, 54, 64, 91
curvature, mean 47
curvature, normal 44
curvature of a curve 13, 15, 17, 18
curvature, principal 45–49
curvature tensor 63, 91
curve, closed 21
curve, constant width 30
curve, convex 27–29
curve, differentiable 106
curve in Euclidean space 8
curve, locally minimizing 100
curve on surfaces 43–45
curve, periodic 21
curve, piecewise smooth 21
curve, pretzel 27
curve, regular 8
curve, simply closed 21
curve, space 17–20
curve, unit speed 9, 43
curve, vertex of 28
cusp 9, 10
cut locus 165
cylinder over a curve 57

Darboux, G. 87, 88, 121, 164, 167
developable surfaces 57–61
Dieudonne, J. 1
diffeomorphism 4
differentiable atlas 105
differentiable curve 106
differentiable function in Euclidean space 3
differentiable function on manifolds 106
differentiable manifold 105
differential 4, 5, 115
differential forms 111–119
directrix of a ruled surface 57
direct sum 111
distance between points in Euclidean space 2
distance between points on manifolds 144
distinguished basis 1
distinguished Frenet frame 11
divergence 76, 91
do Carmo, M. 93, 167
Douglas, J. 71
dual basis 111
dual map 111
dual space 111
Dubins, L. 31
Dupin, C. 55

Edwards, C. 1, 46, 125
Efimov, N. W. 135, 137
eigenvalues as principal curvatures 46
ellipsoid 55, 68, 69
elliptical helix 20
elliptic paraboloid 50, 51
elliptic point 50, 51
embedded submanifold 124
embedded surface 123, 124
embedding into field of geodesics 83
energy 9, 91, 110
equivalence of atlases 105
equivalent sets with Riemannian metric 90
Euclidean space 1
Euler characteristic 141–143
exponential map 96–99
exterior angle 23

Farey, I. 31
Fenchel, W. 31
Fermi coordinates 81
field of geodesics 83
first fundamental form 35–38, 43, 46, 73
Flanders, H. 65, 168
Four vertex theorem 28, 30
Frenet equations 11–15

Frenet frame 10, 11, 78, 91
fundamental group 157
Fundamental theorem of surface theory 64–66

Gauss–Bonnet theorem 138–144
Gauss, C. F. 29, 64
Gauss curvature 30, 47, 48, 54, 64, 91, 92
Gauss equations 62
Gauss frame 35
Gauss lemma 100
Gauss map 35, 38
Gauss's theorem 117, 119
Gauss Theorema Egregium 64
Gauss Theorema Elegantissimum 141
Gauss unit normal field along f 35
generalized cylinder 61
general linear group 127
generated surface 42
generator of ruled surface 57
genus of surface 142
geodesic circles 99, 100
geodesic completeness 60, 154
geodesic curvature 30, 47, 48, 54, 64, 91
geodesic orthogonal coordinates 80–82
geodesic polar coordinates 99–101
geodesics 78–83, 91, 95, 101
geodesics on sphere 79
Gluck, H. 30, 166
gradient 119
graph 127
Green, L. 164
Green's formulae 120
Gromoll, D. 149, 152, 153, 166, 168
Grotemeyer, K. 135
group of transformations 93
Gulliver, R. 72

Hadamard, J. 130, 156
Hadamard's characterization of ovaloids 130
harmonic functions 120
Hartman, P. 61
Heintze, E. 68
Heinz, E. 72
helicoid 42
helix 9, 10
Herglotz, G. 29, 30, 134, 135
Herglotz integral formula 134
Hilbert, D. 68, 93, 167
Hilbert's nonexistence theorem 93
Hildebrandt, S. 71, 72
homotopic curves 154
Hopf, E. 164, 165

Hopf, H. 24, 68, 93, 130, 146, 167
Hopf–Rinow theorem 146
Horn, R. 32
hourglass surface 152
Hurewicz, W. 14, 97
hyperbolic paraboloid 50, 51
hyperbolic plane 92
hyperbolic point 50, 51
hyperboloid 55
hypersphere 126

induced quadratic form 36
inflection point of planar curve 16
injectivity radius 154
injective map 6
inner product 1, 36, 73, 109, 124
integrability conditions 62
integral of differential equation 52
integral of 1-form 117
integral of 2-form 118
interior of a set 2
interior product 115
intrinsic properties 73
invariance of asymptotic directions 53
invariance of curvature tensor 63, 91, 92
invariance of energy integral 91
invariance of first fundamental form 37
invariance of Frenet equations 12, 91
invariance of second fundamental form 39
Inverse function theorem 6
isometric surfaces 83, 84
isometry of Euclidean space 2
isometry 83
isoperimetric inequality 31
isothermal coordinates 120
isotropy subgroups 93
ith curvature of c 13

Jacobian matrix 5
Jacobi, C. G. F. 143, 164
Jacobi fields 102–105, 148
John, F. 52
Jordan curve theorem 31

Karcher, H. 165
kernel 6
Klingenberg, W. 72, 149, 152–154, 163, 168
knotted curve 31
Kobayashi, S. 149, 152, 153, 160, 164, 165, 169
k-times continuously differentiable 4

Laplace-Beltrami operator 120
Lefschetz, S. 138

173

Index

length of curve 9, 91, 110
length of curve on surface 43
length of vector 1
Lewy, H. 162
Lichtenstein, L. 120
Liebmann, H. 137
Lie groups 98, 127
linear functional 111
line element 43, 93
line of curvature 53, 55
Liouville line element 87, 121
Liouville's theorem 122
local diffeomorphism 6
local linearization of differentiable map 6
locally minimizing curve 100, 101
local surface with Riemannian metric 90
Lusternik, L. A. 163

manifolds 89, 105–111
manifolds with Riemannian metric 109
Massey, W. 61, 138, 143, 157
matrix groups 127
matrix representation of bilinear forms 36
mean curvature 47
meridians 87
metric 144
Meusnier's theorem 43, 44
Meyer, W. 149, 152, 153, 166, 168
Milnor, J. 31
Minkowski, H. 136
Minkowski integral formula 136
minimal surfaces 70–72, 121
minimizing curve 100, 101
moving frame along c 10
Myers, S. 165

Nirenberg, L. 61, 162, 168
Nitsche, J. C. C. 70, 71, 121
Nomizu, K. 152, 169
nondegenerate quadratic form 112
non-orientable surfaces 111
norm 1
normal coordinates 99
normal curvature 44
normal form of surface 49–53
normal local representation of space curve 18
normal plane 19
normal section 44
normal vector field 34

1-forms 111, 113, 114
1–1 map 6

one parameter group of isometries 41
onto map 6
open set 6
open surfaces in the large 166
orbit 42
orientable surfaces 110, 128
orientation 110
orientation preserving change of variables 8, 34
orientation preserving motion 2
orthogonal component of isometry 2
orthogonal coordinates 77
orthogonal group 127
orthogonal motion 2
orthogonal transformation 2
osculating circle 45
osculating cone 58, 77, 78
osculating developable 58, 85
osculating plane 19
Osserman, R. 31, 71
ovaloids 129–137

parabolic cylinder 50, 51
parabolic point 50, 51
parallel circles 87
parallel translation 76–78
parallel vector field 76, 77, 85
parameterization by arc length 9
parameterized surface patch 33
parameter transformation of a curve 8
periodic curve 21
piecewise smooth curve 21
plane curves 15–17
planar point 49, 50
planar surface 49
Plateau problem 71, 72
Pogorelov, A. 61, 162, 163
Poincaré, H. 165
Poincaré half plane 92, 94, 95
polar coordinates 99, 100
polygon 116
polygonal decomposition of manifold 118
positive definite quadratic form 35, 133
positively equivalent sets with Riemannian metric 90
positively oriented basis 11
positively oriented manifold 90, 110
pre-geodesic 79
Preissmann, L. 163, 164
pretzel curve 27
prime geodesics 156
principal curvature 45–49
principal curvature coordinates 53
principal directions 45, 46

Index

projective plane 107
pseudosphere 67, 86, 93

quadratic form 35, 112, 133

radius of injectivity 145
Rado, T. 71
realization of surface with Riemannian metric 93
rectifying plane 19
reflection 2
regular curve 8
regular surface 33
Rembs, E. 135
representation of tangent space 108
Riemannian manifold 89
Riemannian metric 30, 90, 109, 110, 124
Riemann normal coordinates 99
rigidity of ovaloids 136
Rinow, W. 146
Rodriguez 46
rotation 2
rotation number 21–23
ruled surfaces 56, 57

scalar product 1
Schnirelmann, L. 163
Schwarz inequality 1
second fundamental form 38, 39, 43, 46
second order surfaces 127
Seifert, H. 138, 164
simplex 116
simply closed curve 21
Singer, D. 166
Singer, I. M. 158, 168
singular polygon 116
singular simplex 116
skew symmetric bilinear form 112
space curves 17–20
special linear group 94
special orthogonal group 127
sphere 40, 66, 77, 86, 106, 126, 137
spherical surface 49
Spivak, M. 1, 51, 65, 127, 168
Spruck, J. 72
star shaped 22
Steiner, J. 31
Stoker, J. 69
straight line 10, 16, 17
Strubecker, K. 67, 70, 168
Sturm comparison theorem 150
Sturm, J. C. F. 150
surface 33, 105, 123
surface of genus g 142

surface patch 33
surface generated by one parameter group of isometries 41
surfaces of revolution 41, 66, 70, 87, 88
surface with Riemannian metric 90
surjective map 6
symmetric bilinear form 35
symmetry 2

tangent bundle 95, 96, 109
tangent bundle of open set 5
tangent field along c 8
tangential developable surface 47
tangential vector field 34, 74
tangential mapping of curve 22
tangent space 96, 108
tangent space of a surface 33
tangent space of Euclidean space 5
tangent vector 95, 96, 108
tangent vector field of c 8
Theorema Egregium 64
Theorema Elegantissimum 141
third fundamental form 39, 48
Thorbergsson, G. 30
Thorpe, J. 158, 168
Threlfall, W. 138, 164
topological atlas 105
topological manifold 105
torsion of space curve 17, 18
torus 40, 126
total curvature 31
tractrix 67
transformation groups 93
transitive group action 93
translation 2
trefoil 27
triangle inequality 1, 144
triangulation 141
triply orthogonal family of surfaces 54
twice continuously differentiable 4
2-form 111, 113, 114

umbilic 49
Umlaufsatz 24
unit normal field along f 35
unit speed curve 9, 43
unparameterized curve 9
unparameterized surface 34

vector field 109
vector field along c 8
vector field along f 34
vertex of a curve 28
Voss, K. 68

Index

Warner, F. 127, 169
wedge product 112
Weingarten map 39, 53
Weingarten surface 68
Weinstein, A. 164
Weyl, H. 161, 162

wiedersehen pair 164
wiedersehen surface 164
W-surface 68

Zoll, O. 164
Zoll surfaces 164

Index of Symbols

$\mathbb{R}^n$, 1
$x \cdot y$, 1
$|x|$, 1
(e_i), 1
$d(x, y)$, 2
$B_\epsilon(x)$, 2
$\mathring{W}$, 2
(a_j^i), 3
$|L|$, 3
$L(\mathbb{R}^n, \mathbb{R}^m)$, 3
$o(x)$, 3
$L(F, x_0)$, 4
dF, 4
$x \times y$, 4
C^1, C^2, 4
C^k, 4
C^∞, 4
d^2F, 5
$T_{x_0}\mathbb{R}^n$, 5
$\mathbb{R}^n_{x_0}$, 5
TU, 5
π, 5
TF_{x_0}, 5
dF_{x_0}, 5
$\ker L$, 6
$c, \dot{c}$, 8, 43
d/dt, 8
$L(c)$, 9, 91, 110
$E(c)$, 9, 91, 110
δ_{ij}, 10
$c^{(k)}(t)$, 10
$\omega_{ij}(t)$, 11

κ_i, 13
κ, 15
$\theta(t)$, 16
τ, 17
S^1, 22
n_ϵ, 23
$K(c)$, 31
$S^{n-1}(r)$, 32
U, 33
Tf_u, 33
D^2, 33
$T_u f$, 33
$df_u e_1$, 33
$f_u 1$, 33
$X(u)$, 34
n, 35
$\beta(X, Y)$, 35
g_{ij}, 36
$I = I_u$, 36
$df \cdot df$, 36
E, F, G, 36
g_u, 36
S^2, 38
$II = II_u$, 39
$-dn \cdot df$, 39
L_u, 39
h_{ij}, 39
L, M, N, 39
$III = III_u$, 39
$dn \cdot dn$, 39
γ_t, 41
$\dot{c}(t)$, 43

Index of Symbols

$|c(t)|^2$, 43
ds^2, 43
$S_u^1 f$, 45
$K(u)$, 47
$H(u)$, 47
$\phi_i, \phi_{,i}$, 61
ϕ_{ik}, 61
g^{ik}, 61
$\Gamma_{ik}^l, \Gamma_{kij}$, 61
R_{ilkj}, 63
$R(X, Y, Z, W)$, 63
pr_u, 74
$\nabla X(t)/dt$, 74
∇X, 76
$\text{div } X(u)$, 76
$\| \cdot \|_c$, 77
$\kappa_g(t)$, 78
$S(2)$, 90
g, 90
Γ_{ij}^k, 91
$\nabla e_i(u)/\partial u^j$, 91
$\nabla X(u)_j^k$, 91
$\text{div } X$, 91
R_{ilkj}, 91
K, 92
H_r^2, 92
$SL(2, \mathbb{R})$, 94
$SO(2)$, 94
TU, 95
$\pi = \pi_u$, 95
$T\phi$, 95
X_u, 95
$T_p M$, 96
TM, 96
g_p, 96
$|X|$, 96
$B_\epsilon M$, 96
$B_\epsilon(0)$, 96
c_X, 96
$\exp$, 97
B_ρ^e, 97
$B_\epsilon U_0$, 97
c_{qr}, 98
$B_e(p)$, 99
$Y(t)$, 102
$S_r^1(p)$, 104
$L(r)$, 104
$A(r)$, 104
$K(p)$, 104

M, 105
$(u_\alpha, M_\alpha)_{\alpha \in A}$, 105
P^2, 107
$T_p M$, 108
TM, 108
π, 108
$g_\alpha(\, , \,)$, 109
T^*, 111
L^*, 111
$T \oplus T$, 111
Ω, 111, 113
$\Lambda^2 T^*$, 112
$\Lambda^2 L^*$, 112
$e^1 \wedge e^2$, 112
$\pi \oplus \pi$, 112
ω, 113
ω_α, 113, 114
$T^* M$, 113
$\Lambda^2 T^* M$, 113
du_α^1, 113
Ω_α, 114
df, 115
L_{g_p}, 115
dM, 115
$i_X dM$, 115
∂F, 116
$\int_c \omega$, 117
$\iint_M \Omega$, 118
Π, 118
$\text{grad } \psi(p)$, 119
$\mathscr{F}(M)$, 120
$\Delta \psi$, 120
$S_r^2(x_0)$, 126
$S_r^{n-1}(x_0)$, 126
$GL(n, \mathbb{R})$, 127
$O(n)$, 127
$SO(n)$, 127
$\text{graph } f$, 127
$\mathscr{H}_p$, 132
$\check{k}^{ij}$, 132
$\chi_\Pi(M)$, 142
T^2, 142
$d(p, q)$, 144
$\beta * \alpha$, 157
$[\beta]$, 157
$\Omega(p)$, 157
$\pi_1(p)$, 157
$\text{Per } TM$, 164

Graduate Texts in Mathematics

Soft and hard cover editions are available for each volume up to vol. 14, hard cover only from Vol. 15

1   TAKEUTI/ZARING. Introduction to Axiomatic Set Theory.
    vii, 250 pages. 1971.
2   OXTOBY. Measure and Category. viii, 95 pages. 1971.
3   SCHAEFFER. Topological Vector Spaces. xi, 294 pages. 1971.
4   HILTON/STAMMBACH. A Course in Homological Algebra.
    ix, 338 pages. 1971. (Hard cover edition only)
5   MACLANE. Categories for the Working Mathematician.
    ix, 262 pages. 1972.
6   HUGHES/PIPER. Projective Planes. xii, 291 pages. 1973.
7   SERRE. A Course in Arithmetic. x, 115 pages. 1973.
8   TAKEUTI/ZARING. Axiomatic Set Theory. viii, 238 pages. 1973.
9   HUMPHREYS. Introduction to Lie Algebras and Representation Theory.
    xiv, 169 pages. 1972.
10 COHEN. A Course in Simple Homotopy Theory. xii, 114 pages. 1973.
11 CONWAY. Functions of One Complex Variable. 2nd corrected reprint.
    xiii, 313 pages. 1975. (Hard cover edition only.)
12 BEALS. Advanced Mathematical Analysis. xi, 230 pages. 1973.
13 ANDERSON/FULLER. Rings and Categories of Modules.
    ix, 339 pages. 1974.
14 GOLUBITSKY/GUILLEMIN. Stable Mappings and Their Singularities.
    x, 211 pages. 1974.
15 BERBERIAN. Lectures in Functional Analysis and Operator Theory.
    x, 356 pages. 1974.
16 WINTER. The Structure of Fields. xiii, 205 pages. 1974.
17 ROSENBLATT. Random Processes. 2nd ed. x, 228 pages. 1974.
18 HALMOS. Measure Theory. xi, 304 pages. 1974.
19 HALMOS. A Hilbert Space Problem Book. xvii, 365 pages. 1974.
20 HUSEMOLLER. Fibre Bundles. 2nd ed. xvi, 344 pages. 1975.
21 HUMPHREYS. Linear Algebraic Groups. xiv. 272 pages. 1975.
22 BARNES/MACK. An Algebraic Introduction to Mathematical Logic.
    x, 137 pages. 1975.
23 GREUB. Linear Algebra. 4th ed. xvii, 451 pages. 1975.
24 HOLMES. Geometric Functional Analysis and Its Applications.
    x, 246 pages. 1975.
25 HEWITT/STROMBERG. Real and Abstract Analysis. 3rd printing.
    viii, 476 pages. 1975.
26 MANES. Algebraic Theories. x, 356 pages. 1976.

27 KELLEY. General Topology. xiv, 298 pages. 1975.
28 ZARISKI/SAMUEL. Commutative Algebra I. xi, 329 pages. 1975.
29 ZARISKI/SAMUEL. Commutative Algebra II. x, 414 pages. 1976.
30 JACOBSON. Lectures in Abstract Alegbra I: Basic Concepts. xii, 205 pages. 1976.
31 JACOBSON. Lectures in Abstract Algebra II: Linear Algebra. xii, 280 pages. 1975.
32 JACOBSON. Lectures in Abstract Algebra III: Theory of Fields and Galois Theory. ix, 324 pages. 1976.
33 HIRSCH. Differential Topology. x, 222 pages. 1976.
34 SPITZER. Principles of Random Walk. 2nd ed. xiii, 408 pages. 1976.
35 WERMER. Banach Algebras and Several Complex Variables. 2nd ed. xiv, 162 pages. 1976.
36 KELLEY/NAMIOKA. Linear Topological Spaces. xv, 256 pages. 1976.
37 MONK. Mathematical Logic. x, 531 pages. 1976.
38 GRAUERT/FRITZSCHE. Several Complex Variables. viii, 207 pages. 1976
39 ARVESON. An Invitation to C^*-Algebras. x, 106 pages. 1976.
40 KEMENY/SNELL/KNAPP. Denumerable Markov Chains. 2nd ed. xii, 484 pages. 1976.
41 APOSTOL. Modular Functions and Dirichlet Series in Number Theory. x, 198 pages. 1976.
42 SERRE. Linear Representations of Finite Groups. 176 pages. 1977.
43 GILLMAN/JERISON. Rings of Continuous Functions. xiii, 300 pages. 1976.
44 KENDIG. Elementary Algebraic Geometry. viii, 309 pages. 1977.
45 LOEVE. Probability Theory. 4th ed. Vol. 1. xvii, 425 pages. 1977.
46 LOEVE. Probability Theory. 4th ed. Vol. 2. approx. 350 pages. 1977.
47 MOISE. Geometric Topology in Dimensions 2 and 3. x, 262 pages. 1977.
48 SACHS/WU. General Relativity for Mathematicians. xii, 291 pages. 1977.
49 GRUENBERG/WEIR. Linear Geometry. 2nd ed. x, 198 pages. 1977.
50 EDWARDS. Fermat's Last Theorem. xv, 410 pages. 1977.
51 KLINGENBERG. A Course in Differential Geometry. xii, 192 pages. 1978.
52 HARTSHORNE. Algebraic Geometry. xvi, 496 pages. 1977.
53 MANIN. A Course in Mathematical Logic. xiii, 286 pages. 1977.
54 GRAVER/WATKINS. Combinatorics with Emphasis on the Theory of Graphs. xv, 368 pages. 1977.
55 BROWN/PEARCY. Introduction to Operator Theory. Vol. 1: Elements of Functional Analysis. xiv, 474 pages. 1977.
56 MASSEY. Algebraic Topology: An Introduction. xxi, 261 pages. 1977.
57 CROWELL/FOX. Introduction to Knot Theory. x, 182 pages. 1977.
58 KOBLITZ. p-adic Numbers, p-adic Analysis, and Zeta-Functions. x, 122 pages. 1977.